The New World of Russian Small Arms & Ammo

The New World of Russian Small Arms & Ammo

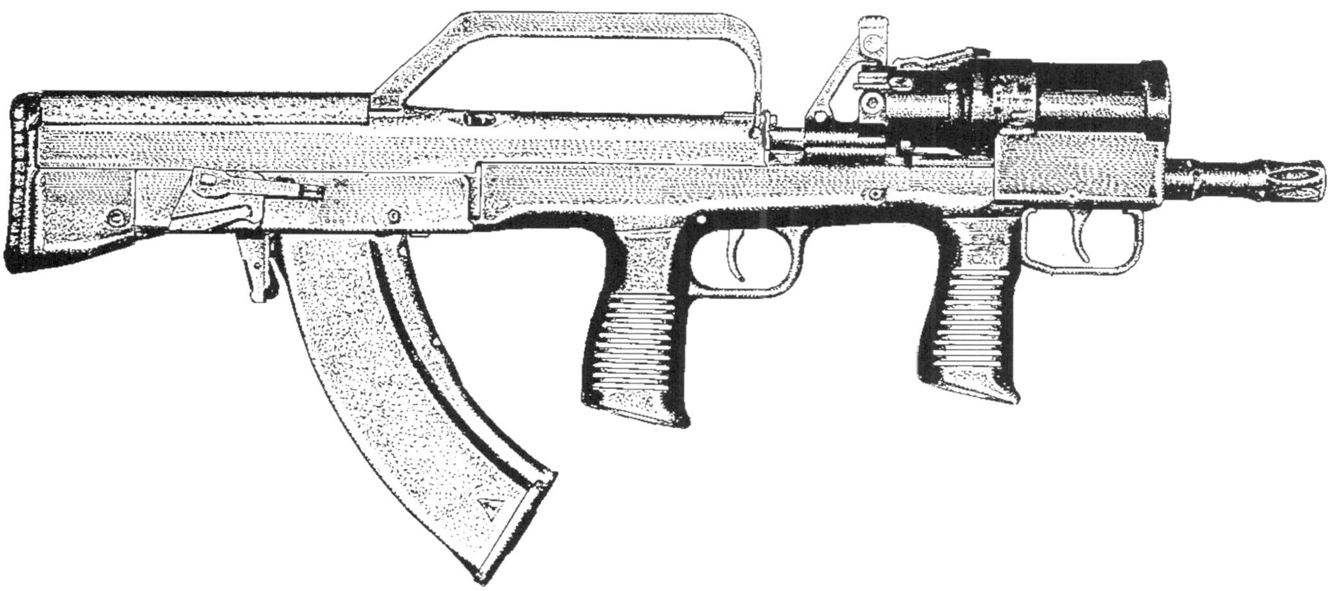

Charlie Cutshaw

Illustrations by Lyn Haywood

Paladin Press · Boulder, Colorado

The New World of Russian Small Arms and Ammo
by Charlie Cutshaw
Illustrations by Lyn Haywood

ISBN 0-87364-993-1
Printed in the United States of America

Published by Paladin Press, a division of
Paladin Enterprises, Inc., P.O. Box 1307,
Boulder, Colorado 80306, USA.
(303) 443-7250

Direct inquiries and/or orders to the above address.

CONTENTS

INTRODUCTION

When Russian small arms come to mind, one usually thinks of the ubiquitous Kalashnikov AK-47 or one of its successors that have been the mainstay of Soviet and Russian small arms since 1949. In recent years, especially in the United States, millions of Simonov (SKS) carbines have been sold commercially, and the little rifle is familiar to anyone who reads the firearms press. If one's interest went beyond SKS carbines, AK-type assault rifles, certain machine guns, and such pistols as the TT-33, PM Makarov, and APS Stechkin, little else was known. There was nothing available in the West regarding new and developmental firearms, much less details on new ammunition both in traditional and new calibers.

With the demise of the Soviet Union, however, things changed. It soon became obvious that Russian small-arms designers had not been idle under Soviet rule, but had developed, and in several instances fielded, some of the most innovative small arms in history. Included were underwater weapons; weapons that used silent ammunition; unknown families of submachine guns, pistols, and revolvers that demonstrated truly advanced thinking and design; plus innovative new assault rifles, one of which appears destined to replace the Kalashnikov in Russian service. During the same period, Russian designers also developed a variety of new types of ammunition to accompany their weapons.

What is remarkable is that several of these innovative weapons have been in Soviet military service since the early 1970s, completely unknown to the West or any of its intelligence services. One, at least, was actually used operationally in Afghanistan and Central America. Why this never came to light will become apparent to the reader upon perusal of the section on that peculiar weapon. The lack of knowledge concerning these weapons should come as no great surprise, as most of them were designed for use by special operations forces—SPETsNAZ, or special direction forces—whose use of them was understandably kept classified so that Western adversaries would be kept unaware of their potential opponents' armament. Despite their secrecy, there were usually hints in the West of new Soviet weapons development, but the existence of these remarkable small arms was successfully kept under wraps until the Russians chose to release information on them to sell them overseas to raise capital. Even so, there is still a sense of secrecy on the part of the Russian arms industry. For example, as of this writing, certain aspects of the new AN-94 assault rifle are still classified as state secrets, although the rifle has been seen and examined by several Westerners. The same is true with many of the other weapons described in this book; although the Russian small-arms industry desperately needs capital, its members are reluctant to divulge any more information than absolutely necessary to members of the press, authors, or visitors from Western small-arms manufacturers. The reader will, therefore, find a plethora of information on some weapons, but little more than brief outlines on others.

It is also difficult to differentiate between weapons that are merely prototypes or "trial balloons" for potential joint projects with Western manufacturers and weapons that are actually in service with Russian military or interior ministry forces. When asked, representatives of Russian arms makers will almost invariably state that the firearm in question is in use by the army or Ministry of Internal Affairs (MVD) or SPETsNAZ, yet personnel of these organizations may never be observed with them. This does not mean that a weapon in question is not in use as claimed; only that it has never been observed. Nonetheless, true confirmation of use comes only when a weapon is actually observed in the hands of operational forces. I will make every attempt in this volume to differentiate and categorize weapons.

New weapons are the subject of this book, although I will discuss the latest and probably final Russian versions of the venerable Kalashnikov series assault rifles. The Russian military's new assault rifle, the AN-94, has created a great deal of interest in the West, but little has been written about it outside the military small-arms community. This significant new weapon will be covered in as much detail as possible. I will also discuss several prototype weapons that never entered production, but which are important for the contribution they made to overall weapons design. The final portion of this book will cover Russian ammunition developments, which are every bit as fascinating and innovative as the firearms that use them.

A book such as this one is the product of many people besides the author whose name appears on the title page. First, I must thank Dave Markov, who first suggested the project and who provided much of the data that made this book possible. I also wish to thank my friend and colleague Valery Shilin without whose assistance this book would never have been written. I also owe a deep debt to Cookie Sewell and Les Grau, who willingly translated many Russian-language documents for me at no cost. Again, I couldn't have done this without these gentlemen. Maj. Marc Moo Sang of the Canadian army is yet another without whose assistance this book would not have been possible. I owe Col. Rex Applegate a word of thanks for encouraging the project and using his good offices to help get the book published. Don Wood is yet another gentleman whose generous assistance made this book possible. My friend and colleague Terry Gander also gave inestimable help. I'd also like to thank my editor at Paladin, Donna DuVall, whose efforts and competence made the publication process painless for me. (I suspect that it was not so for her.) And finally, I must thank my wife, Dianne, whose encouragement and confidence in my abilities made possible not only this book but everything else.

CHAPTER ONE

RIFLES

AN-94 ASSAULT RIFLE

In 1993, a mysterious new assault rifle appeared at an arms display at the elite Taman Guards Division outside Moscow. The strange new rifle was displayed alongside the common AK-74 assault rifles and RPK-74 light machine guns that made up the standard light-weapons armament of the division and was labeled "ASN," which was subsequently learned to be a Russian abbreviation for *Avtomat Spetsialnyi Nikonov*.

Soldiers at the show could give no meaningful information on the weapon other than the basic data on the placard above the rifle displayed on the table. Shortly thereafter, more information began to emerge regarding the ASN, indicating that it was a developmental advanced combat rifle in the true sense of the word, but its official status remained a mystery. The Russian Ministry of Defense cleared up the mystery in 1996, however, by announcing that the ASN had passed all its troop trials and had officially been type-classified as the Avtomat Nikonova-94 (AN-94), named for the head of the team that designed it, Gennady Nikonov. The number 94 indicates the year that the rifle was officially type-classified and adopted for military service to eventually replace the AK-47/AKM/AK-74 series rifles.

Rumors from Russia had it that the "grand old man" of

Russian small arms, Mikhail Kalashnikov, was not pleased by the new rifle, especially since a design team headed by his son had been contending for the honor of designing Russia's next assault rifle. Nonetheless, as time passed, it was clear that the AN-94 would almost certainly be the next Russian combat rifle, Kalashnikov's protestations notwithstanding.

The genesis of the AN-94 actually begins with the adoption of the AK-74 more than 20 years ago. At the time, the 5.45x39mm cartridge of the AK-74 was tacit acknowledgment by the Soviet military of the combat effectiveness of small-caliber high-velocity projectiles at normal battlefield ranges that had been proven in Vietnam by the 5.56x45mm M16A1. The new cartridge was derived from the M1943 7.62x39mm cartridge. Although the 5.45mm's terminal effects were sufficiently lethal to earn it the nickname "the poison bullet" by the Mujahideen in Afghanistan, the Soviet military was not completely satisfied with the overall performance of the AK-74. The military had probably concluded that the AK-74 effectively represented the end of the practical development life of the Kalashnikov assault rifle design and that a new rifle would be necessary by the turn of the century.

The primary requirement of a new rifle was that it would achieve a probability of hit (or effectiveness) of 1.5 to 2.0 times that of the AK-74. The military had also apparently felt that a reduction in recoil was necessary to improve hit probability—ironically, a consideration in light of the fact that reducing recoil and improving control and hit probability had been part of the reason for the development of the AK-74. The AK-74 had reduced recoil in comparison with the older 7.62x39mm AK-47s and AKMs, but the reduction was due to the ammunition change and, to a lesser extent, the extremely effective AK-74 compensator/flash suppressor design. However, the reduction was considered inadequate, and a program was initiated to develop a new, advanced-technology assault rifle to replace the entire Kalashnikov family. In addition to increased effectiveness and reduced recoil, the new rifle would also have to meet stringent increased reliability requirements.

The development program was nicknamed *Abakan* for a village in Siberia where a great deal of the testing of the candidate rifles was conducted. This led to the AN-94's being mistakenly called *Abakan* when it was first observed in the hands of Russian troops several years ago. *Abakan* was also used by the Russians as a generic term for all rifles that were competing to be the next Russian combat rifle, further confusing the issue. According to Gennady Nikonov, designer of the AN-94, every official Soviet firearms designer submitted a candidate for consider-

The 5.45mm ASN. This 1980s design was contemporary with the Abakan design by G.H. Nikonov. It was followed by two designs, the AS and the ASN, the latter completed in 1987.

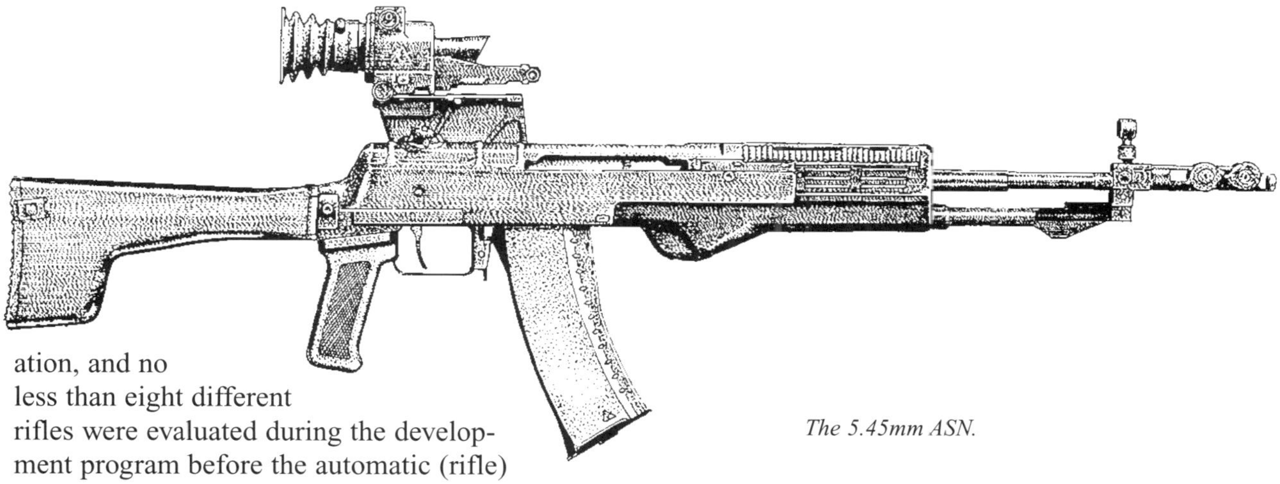

The 5.45mm ASN.

ation, and no less than eight different rifles were evaluated during the development program before the automatic (rifle) system Nikonova (ASN) was eventually selected and type-classified as the AN-94.

Interestingly, in 1992 the usually plain-spoken Mikhail Kalashnikov refused to comment on the ASN, then undergoing troop trials. He was quoted as saying, "I don't feel I am entitled to give an assessment of the new product." Kalashnikov went on to state that his son Viktor had a design in the Abakan competition, hence his reluctance to comment. The senior Kalashnikov's reluctance to comment apparently disappeared after Viktor's candidate lost. No direct quotes have been made public, but informal information from Russia was to the effect that the elder Kalashnikov was furious that his son's rifle had lost and that Kalashnikov senior did everything in his power to reverse the decision. It is thus apparent that the Kalashnikov design was a "loser" in the competition, especially given Kalashnikov's protestations about adoption of the AN-94. Viktor Kalashnikov's design has yet to be made public.

As of early 1998, the AN-94 was in limited production at the Izmash Joint Stock Company and is truly a complete departure from earlier Soviet/Russian designs. For that matter, it can truly be stated that the AN-94 is unique. The weapon shares only five components with the AK-74: the magazine, folding stock hinge, pistol grip, optical sight base, and cleaning kit. Full production of the AN-94 has been delayed by lack of funds and perhaps by political efforts of Mikhail Kalashnikov to have his son's rifle replace the AN-94. Despite Kalashnikov's efforts, authoritative Russian sources have stated that the AN-94 will eventually enter Russian military service and replace the AK series. The AN-94 can essentially be considered as a Russian version of the U.S. advanced combat rifle—perhaps even the objective individual combat weapon without all the bells and whistles of the latter. This becomes apparent when the capabilities of the AN-94 are compared with any other combat rifle in the world, either in service or in development.

It is possible that the Kalashnikov-derived 6x49mm rifle and machine gun, which were shown by the Precision Mechanical Engineering Central Research Institute (TsNIITochmash) at numerous international arms shows in the mid-1990s for potential joint production with overseas partners, were among the losers in the

Abakan competition. Losing in the Abakan competition was probably due to the Kalashnikov weapons' being no better in performance than the AN-94 and their having to use a new cartridge to achieve their proclaimed superiority.

Although the Kalashnikov derivatives would have led to simplified training because of their being based on existing designs, they would have placed a tremendous, probably unacceptable, demand on an ammunition program that already has three rifle and light/general-purpose machine gun calibers. The advantage of the AN-94 over the Kalashnikov designs is that it achieves the same or better performance while simplifying logistics by using existing ammunition, not to mention claimed AN-94 reliability improvements over the Kalashnikov rifles. Moreover, the capabilities and overall design of the AN-94 indicate that the Russian military has undergone a change in its small-unit tactical thinking, as will be elaborated upon.

Construction methods of the AN-94 were likely another reason for its adoption over its competitors. Unlike the Kalashnikov designs, the AN-94 is a far more modern rifle, using state-of-the-art manufacturing techniques. The AN-94's furniture is completely manufactured of polymer, as is that of the AK-100 series, but there the similarity between the two weapons ends. What appears to be a gas tube beneath the barrel of the AN-94 is actually a fixed rod extending from the stock group. This rod incorporates a guide for the rifle's firing unit (barrel-receiver assembly) at the front and a dual-purpose stud at its center. One purpose of the stud is to stabilize the rifle when it is being fired in fully automatic mode. The barrel and gas tube of any fully automatic firearm tend to resonate, and the stud counters this tendency, thereby increasing accuracy. A second purpose of the stud is to prevent the rifle from sliding out of place while it is engaged in the firing port of an infantry fighting vehicle. The actual gas cylinder of the AN-94 is located above the barrel and underneath the handguard and is quite short. The reason for this will become apparent as I discuss the functioning of the AN-94.

The entire operating mechanism of the AN-94 is inside the stock, referred to in Russian documents as a "carrier-stock." The reason for this is that the barrel and

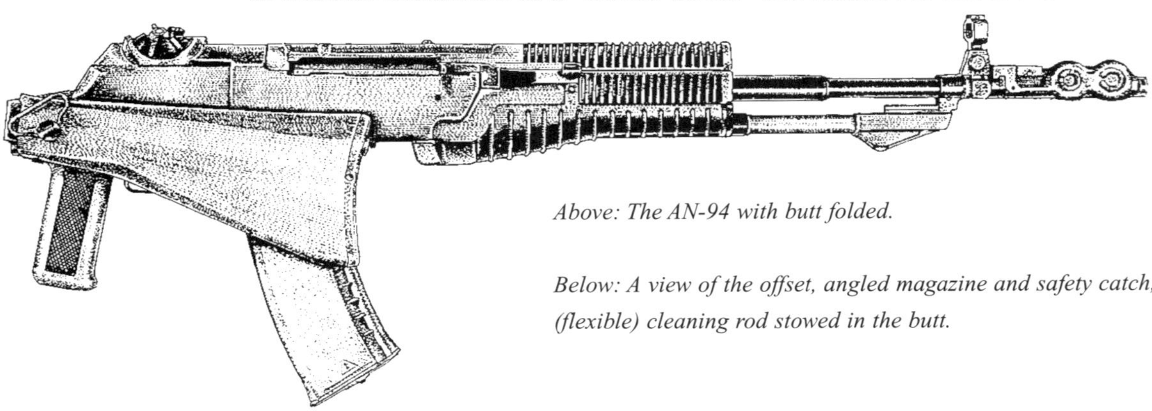

Above: The AN-94 with butt folded.

Below: A view of the offset, angled magazine and safety catch, and the (flexible) cleaning rod stowed in the butt.

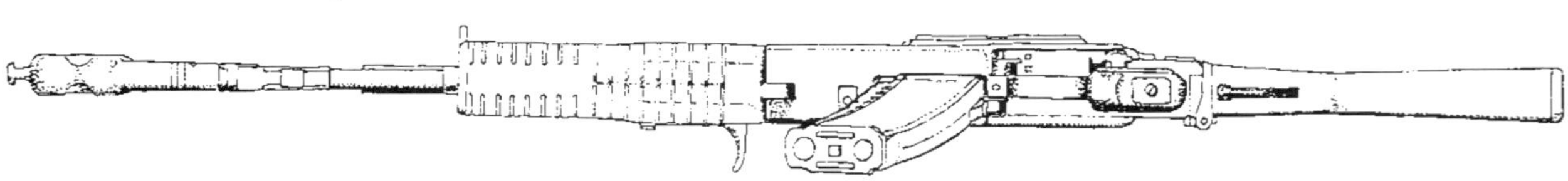

receiver are integrated into a firing unit that reciprocates on guide rails inside the stock. The bolt carrier and bolt are carried by and operate inside the rifle's internal receiver. The AN-94 has two internal buffers, one in the forearm and another in the rear of the receiver. The forearm buffer not only absorbs shock, but also accelerates return travel of the firing unit as it moves forward in counterrecoil. The rear buffer boosts forward acceleration of the bolt carrier and also prevents the receiver-firing unit from striking the rear of the stock as it recoils. The magazine, which is interchangeable with those of the AK-74 and RPK-74, is offset to the right and inserted at a slight angle.

One of the key principles of the AN-94's operation is referred to as a *blowback shifted pulse*, or BBSP. This stems from the fact that the receiver-barrel assembly reciprocates independently from the bolt and its carrier, although the latter reciprocate in the receiver-barrel assembly. A simplified functional explanation of the two-round burst feature of the AN-94 will clarify the BBSP principle.

When the first round is fired, the entire barrel-receiver assembly begins moving to the rear under recoil forces, taking the bolt carrier with it and compressing the forward buffer. The bolt is still locked to the barrel at this point. As the bullet passes the gas port, gas is bled off into the gas cylinder, driving the bolt carrier to the rear inside the receiver-barrel assembly, and at the same time unlocking the bolt and extracting and ejecting the spent cartridge case. The bolt carrier is moving much faster than the barrel-receiver assembly and strikes against the rear buffer, which in conjunction with the return spring propels it forward, temporarily inactivating the sear while stripping a fresh round from the magazine and chambering it. As the bolt locks, the sear is released and the second round is fired before the receiver completes its rearward motion. In essence, the first two bullets have left the barrel of the rifle while the receiver is still moving to the rear and has not had a chance to hit the rear buffer. The cyclic rate of this two-round burst feature is 1,800 rounds per minute (RPM). As can be seen, the AN-94 is both gas and recoil operated. Although this complex yet robust BBSP system is not all that makes the AN-94 unique, it is the central principle that makes the dual cyclic rate possible, and this capability is in turn what gives the AN-94 its remarkable performance improvements over all currently produced assault rifles.

The functioning of the AN-94 is unique. When set for fully automatic fire, the first two rounds out of the AN-94's barrel will always be at the "high" rate of 1,800 RPM, so fast that the bullets will be on their way downrange before the rifle has a chance to recoil in the shooter's hands, thereby increasing the probability of a hit over the entire effective range of the rifle. Once the first two rounds are on their way, the rifle automatically cycles down to 600 RPM. When the trigger is released, the mechanism resets for 1,800-RPM two-round burst fire.

To say that the AN-94 is "different" from any currently produced combat rifle in the world is an understatement. Not only are the capabilities unique, but so is the method used to achieve them.

The AN-94's functioning can best be understood in the context of the small arms cycle of operation. It is difficult to describe the functioning of a firearm without having examined it; the technical description of the AN-94's functioning is based solely on data provided to me by its manufacturer and is subject to my judgment in interpreting them. When and if an actual AN-94 is placed in my hands for examination, the following description may have to be revised. Because of length considerations, only the fully automatic cycle of the AN-94 will be described, because this captures all of the rifle's unique operating features.

Firing the AN-94

To fire, the (separate) safety is set on *O*, or *ogon* (fire in Russian), and the selector switch is set to *AB*, an abbreviation for *avtomatichesky* (automatic). This shifts the disconnector into contact with one of the shoulders of the trigger plate.

Feeding and Chambering

The operator pulls the operating handle to the rear. As he does so, a mechanism consisting of a pulley and cable carries out preliminary feeding, stripping a round from the magazine and placing it in line with the bolt. This operation occurs each time the bolt carrier moves to the rear. It is essentially a "half-loading" of the rifle. The cable assembly also limits rearward travel of the bolt carrier, preventing it from overrunning the magazine. When the operating handle is released, pressure of the return spring and rear buffer drive the bolt carrier forward, ramming the cartridge into the chamber. As can be seen, feeding is a two-stage process, although the operating handle is pulled to the rear only once.

Firing, First Round

Pulling the trigger moves the trigger plate on its axis, releasing the sear, which in turn releases the striker and fires the cartridge.

Unlocking, First Round

Driven by recoil, the barrel-receiver assembly (firing unit) moves to the rear on its guide rails inside the carrier-stock and begins to compress the forward buffer. The bullet passes the gas port, allowing gas into the cylinder, pressing against the gas piston and driving the bolt carrier to the rear inside the moving receiver. The bolt is cammed by the carrier to turn and unlock from the barrel extension. Again, it should be noted that the barrel-receiver unit is moving to the rear as this operation takes place.

Extracting, First Round

As the bolt continues to the rear, it pulls the spent cartridge case from the rifle chamber.

Ejecting, First Round

As the spent cartridge case clears the base of the chamber, it is thrown clear of the rifle out the ejection port.

Second Round, Feeding, Chambering, and Locking

As the bolt carrier and striker reach their rearmost position, the return spring and buffer press them back forward to complete feeding and then chamber and lock the second round for firing at the high cyclic rate. The barrel-receiver is still moving to the rear. Since this unit is further to the rear than when the first round was fired, the second cartridge has a shorter distance to travel from the magazine to the chamber, thereby assisting in achieving the 1,800-RPM rate of fire. This action is further assisted by the "pre-feeding" of each cartridge as the bolt carrier moves to the rear. As the barrel-receiver unit moves to the rear, the sear is temporarily deactivated and the striker freed to fire the second cartridge. Whether the striker follows the bolt forward or is released by the sear as the bolt locks is not clear at the time of this writing. Regardless, the second round fires while the barrel-receiver assembly is still moving rearward. This completes the high cyclic rate.

Third Round, Low Rate

The bolt carrier and bolt again begin to move rearward, driven by gas from the second cartridge. The sear is retained by the trigger plate. The disconnector and trigger plate return to their original positions. The rear buffer and return spring drive the bolt carrier forward again, ramming a preloaded round into the chamber. The barrel-receiver unit is driven forward by the buffers and the return spring. The disconnector moves forward with the barrel-receiver unit after the third round is chambered and the rifle locked, releasing the striker and firing the rifle. Further details of the operation of the AN-94 have not been released by the manufacturer.

Izmash, the Russian manufacturer of the AN-94, claims that the probability of a first-shot hit with the AN-94 is 1.5 to 1.7 times better than the AK-74 by actual troop tests in combat units. Russian sources have also stated that the overall efficiency of the AN-94 is "twice that of the AK-74 and 50 percent greater than the American M16." Whether the M16 referred to was the M16A1 or M16A2 is not clear; the M16A2 is a significant improvement over its predecessor in terms of overall performance. Regardless, it is clear that the AN-94 has achieved the design requirements of the Russian military.

The AN-94 can be equipped with an optical sight, the familiar 1L29 4x optical sight used on the AK-74, RPK-74, and PKM weapons for several years. The AN-94 also has unique "iron" sights. The rear adjustable peep sight is a "wheel" canted slightly to the right of horizontal with apertures at different elevations for

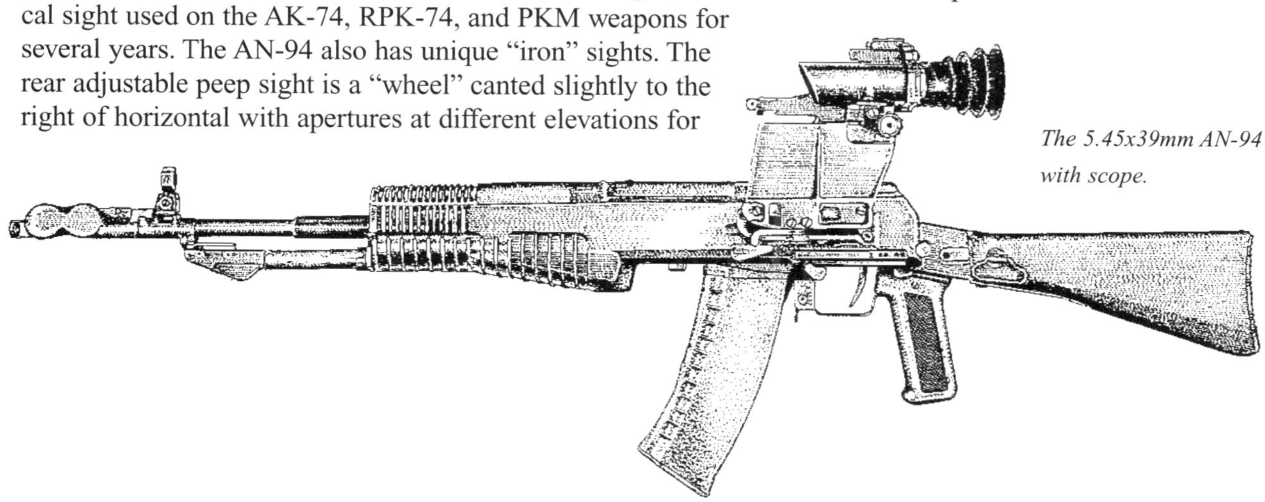

The 5.45x39mm AN-94 with scope.

adjusting aimed fire. According to Russian sources, the AN-94 is sighted out to 1,000 meters, but increments of sight adjustment are unknown at the time of this writing, and 1,000-meter accuracy and effectiveness of the 7N6 52.5-grain or 7N10 55.5-grain bullets are questionable at best. The "canted drum" diopter rear sight is similar in concept to that used by Heckler & Koch, among others. Each aperture is clearly marked on top with its sighted range, and a quick twist by the rifleman allows him to change his range almost instantly, even when wearing gloves or trigger-finger mittens. The disadvantage of the AN-94 system is the inability to make precise changes in elevation, as can be done with M16A2 sights, but the AN-94 sights are well suited for use on a combat rifle. Windage is changed by moving the front sight side to side.

In addition to using the same optics as other Russian weapons, the AN-94 also accommodates the GP-25/GP-30 40mm underbarrel grenade launcher. The bayonet for the AN-94 is similar to older designs, but those seen lack the "saw-back" feature of earlier bayonets, and the point is tapered. The AN-94 bayonet retains the wire-cutting ability of earlier bayonets. Unlike older bayonets, however, the AN-94 bayonet mounts on the right side of the barrel, rather than underneath.

The AN-94 probably reflects a change in Russian infantry tactical thinking. (Russian sources have informally confirmed this fact in personal conversations, but there has been no official announcement, so I have retained the word *probably* in the text.)

There are a number of features of the AN-94 that support this notion. The burst-fire feature and emphasis on accuracy are clear indicators that accurate aimed fire has worked its way into Russian military thinking. All AK-series weapons place an emphasis on full-automatic capability because massed automatic fire was paramount in Soviet infantry tactics at the time of the AK's design in the late 1940s. Accurate aimed fire was a secondary consideration. The AN-94's adjustable, relatively sophisticated, open sights are also a clear indication of a departure in Russian tactical small-arms thinking. Unlike the rudimentary "notch" sights of the AK-series weapons, the open sights of the AN-94 are, as described above, much more sophisticated, lending to accurate placement of fire on the target. Finally, the 1,800-RPM, two-round burst feature of the AN-94 is specifically intended to raise the probability of a hit and increase the effective range of the rifle. These requirements are the antithesis of fully automatic massed fire, which achieves hits by sheer numbers of rounds fired toward the target.

According to Russian sources, the AN-94 is disassembled by "traditional methods," but reliability and maintenance are claimed to be greatly improved over even the latest versions of the AK-series weapons by the aforementioned use of modern materials and production processes. It is difficult to imagine that AK reliability could be improved upon, given the reputation of the Kalashnikov weapons for ruggedness, but the Russians claim that the AN-94's mean number of rounds between failure is 40,000, a 150 percent improvement over the AK-74!

In sum, Gennady Nikonov has apparently designed a thoroughly modern replacement for the venerable Kalashnikov series of assault rifles, which have dominated not only Soviet/Russian but also the world's military small-arms market for nearly 50 years. There are other significant military rifles, but none can compare to the AK series in terms of either numbers, technical qualities, or historical influence. Since the introduction of the AK-47 in 1949, there have been some 50 million AK-type weapons manufactured worldwide. The AK defined the term *assault rifle* and has seen action in virtually every war since the early 1950s. It has even been used on a limited basis by some U.S. military forces. The Kalashnikov continues today as the AK-100 series, which will be discussed in a separate section of this book. These latest versions of the AK design are available in not only traditional Russian chamberings, but in North Atlantic Treaty Organization (NATO) calibers as well. So the AN-94 will never totally supplant its predecessor, at least not for many years. The Russians have wisely begun to replace their basic infantry weapon while the earlier design is still viable; it will take many years to completely replace the AK inventory in the Russian military, and the AK will therefore remain a viable combat rifle well into the 21st century.

AN-94 SPECIFICATIONS

Caliber	5.45x39mm
Operation	Gas, Select Fire
Weight	8.47 lbs. (3.8 kg.)
Length, Stock Extended	37.1 in. (943mm)
Stock Folded	28.6 in. (728mm)
Barrel Length	15.9 in. (405mm)
Effective Range	600 m.
Cyclic Rate	1,800 & 600 RPM
Magazine Capacity	30 & 45 Rounds
Sights	Front, Blade; Rear, Aperture, Adjustable

TULA KBP 7.62X39mm A-91M BULLPUP ASSAULT RIFLE

A real surprise came in early 1998 when Tula Instrument Design Bureau (KBP) announced that it had a new assault rifle available for production. In discussions with me, Tula representatives were reluctant to discuss details regarding the rifle, other than the fact that it is not actually in full production. Nonetheless, the A-91M is a very innovative new rifle that appears to be among the best of the new bullpup rifles that many firearms manufacturers are currently producing. Many of these rifles are based on the Kalashnikov, e.g., the recently announced South African CR-21, the Armenian K-3, the Chinese Type 86, and the Tula OTs-14 *Groza*. Despite grandiose claims, all these rifles are in essence no more than Kalashnikov "spinoffs" modified to bullpup configuration.

The A-91, however, is quite different from any other bullpup, despite its being manufactured by the same Russian firm that produces the OTs-14 *Groza*, a Kalashnikov derivative. At first, the A-91M itself appears to be just another Kalashnikov modified into bullpup configuration, but upon examination, it becomes clear that this is not just another Kalashnikov in bullpup clothing. There is no apparent ejection port, just a small oval hole at the right rear of the carrying handle. This small aperture is indeed the ejection port: spent cases are ejected forward, not to the side as in more conventional firearms. Other than that small aperture, the receiver is completely sealed once the magazine is inserted. The flash suppressor is a Western style "birdcage," rather than the unusual suppressors usually seen on Russian firearms. It is also clear that the gas system is different from that of the Kalashnikov. All furniture is polymer. The nonreciprocating charging handle is located at the left front of the carrying handle and can be pivoted up or down by the shooter. The selector lever is located in the traditional Kalashnikov position, but given the fact that the A-91M is claimed to be fully ambidextrous, there is probably a mirror image selector on the left side of the receiver. The rifle is designed so that the recoil forces are directed to a

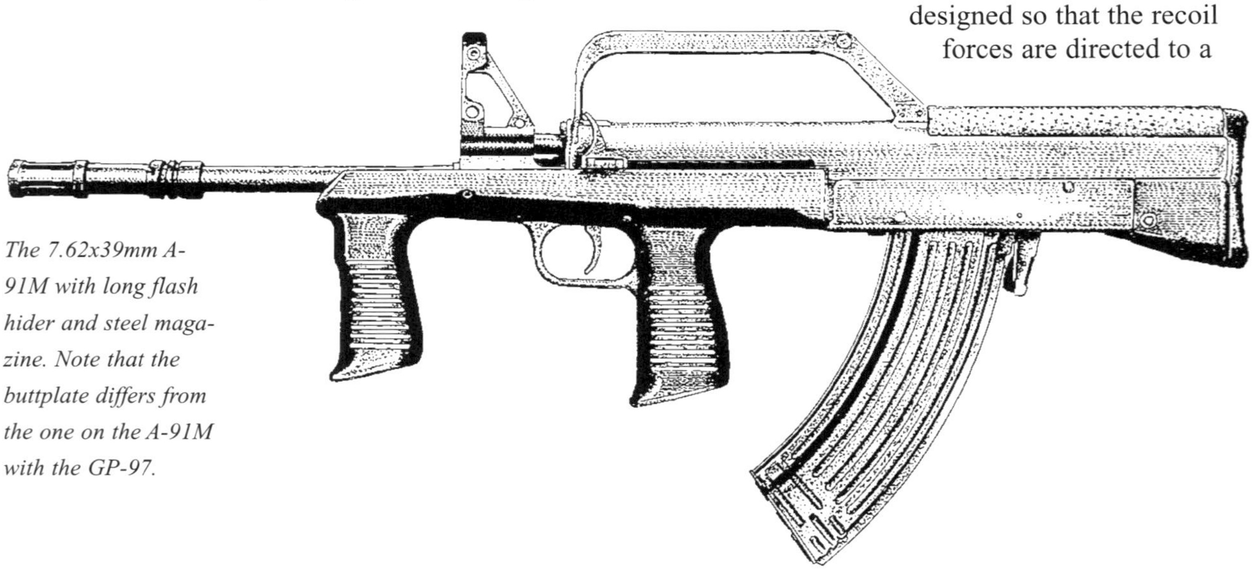

The 7.62x39mm A-91M with long flash hider and steel magazine. Note that the buttplate differs from the one on the A-91M with the GP-97.

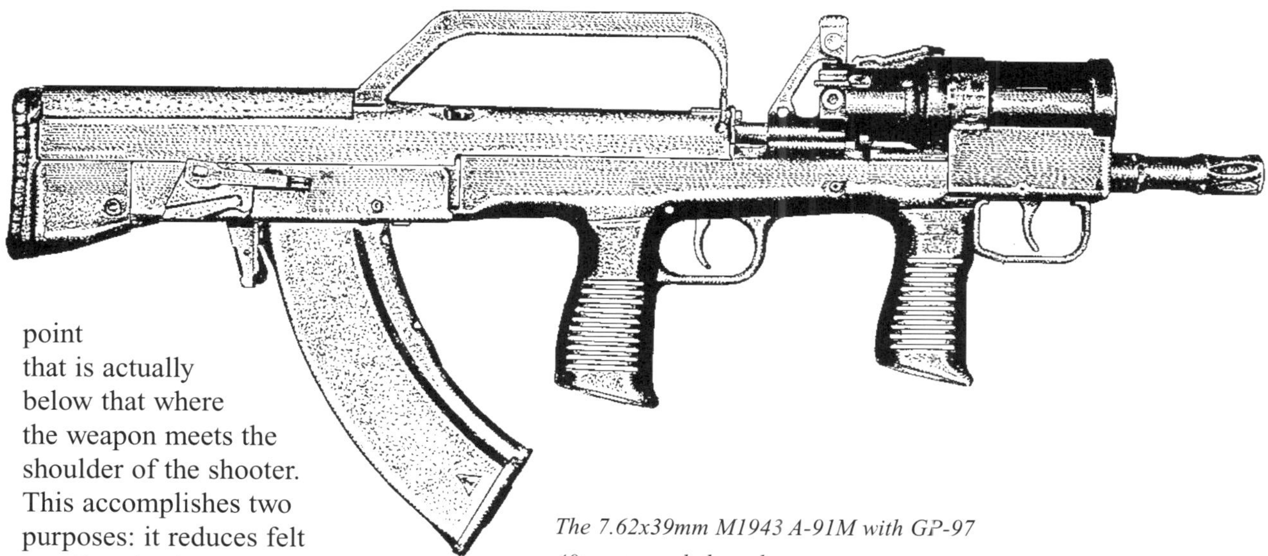

point that is actually below that where the weapon meets the shoulder of the shooter. This accomplishes two purposes: it reduces felt recoil and enhances the ability of the shooter to control the rifle when firing fully automatic. The only component from the Kalashnikov that appears to be shared by the A-91M is the magazine.

The 7.62x39mm M1943 A-91M with GP-97 40mm grenade launcher.

Although the new rifle is designated A-91M, indicating that it is derived from the A-91 compact assault rifle, the functioning and design of the A-91M appear to be either completely different from the A-91 or very extensively modified in the transition from the original rifle's conventional configuration to a bullpup design. It should be noted, for example, that the A-91 is a conventional gas-operated rifle that ejects to the right from a conventional ejection port and has a reciprocating charging handle, while the A-91M functions in quite a different fashion. Because it only recently came to light and there is presently a lack of information regarding it, a full explanation of the role and detailed technical aspects of the A-91M rifle will have to wait. Perhaps it was a loser in the Abakan competition, which resulted in the AN-94. Whatever, the A-91M appears to be a well-executed and fully developed system.

The A-91M's barrel-mounted grenade launcher is designated the GP-97, and it appears to be an upside-down-mounted GP-95 (see Chapter 5), but this has yet to be confirmed. The GP-97 probably mounts onto the rings visible midway between the rear of the flash suppressor and the fore-end. The GP-97's trigger mechanism is ergonomically positioned so that the trigger finger of the support hand can easily access it. The GP-97 fires standard VOG-25 and VOG-25P 40mm grenades. The flash suppressor of the A-91M will also accept muzzle-launched projected grenades. Whether the muzzle is standard NATO 22mm diameter has not been stated.

Why Tula chose to retain the 7.62x39mm cartridge rather than chambering the new rifle in 5.45x39mm, 9x39mm, or even 5.56x45mm as it has done with other rifles is another mystery, particularly considering the Russian military's desire to reduce recoil. Any rifle chambered for the 7.62x39mm cartridge can easily be converted to fire any of the others mentioned. If Tula is serious about marketing this innovative rifle, it will be chambered in alternative calibers in the very near future.

A-91M SPECIFICATIONS

Caliber	7.62x39mm
Operation	Gas, Select Fire
Weight	6.6 lbs. (3 kg.)
Length	23.4 in. (595mm)
Barrel Length	15.7 in. (400mm)
Effective Range	800 m.
Cyclic Rate	600–800 RPM
Magazine Capacity	30 Rounds
Sights	Front, Protected Blade; Rear, Protected Notch (Estimate)

KALASHNIKOV AK-100 SERIES AND OTHER KALASHNIKOV DERIVATIVES

Background

Although the subject of this book is new Russian small arms, the Kalashnikov AK-100 weapons are included because they are relatively unknown in the West and because they are actually new versions with significant changes from their predecessors. But before we delve into the subject, it is appropriate to briefly examine the history of the AK weapons family, setting aside a few myths and misconceptions surrounding the weapons.

Mikhail Timofeyevich Kalashnikov is reputed to have designed the AK-47 while recuperating from battle wounds. Although this story certainly adds to the Kalashnikov mystique, it is false. Kalashnikov was wounded in 1941 and while recuperating did dream of Soviet troops armed with thousands of submachine guns attacking the hated Germans. After his release from the hospital, Kalashnikov was sent to work in a political office. It was here that he designed his first weapon, a submachine gun designed around the TT-33 cartridge. Kalashnikov got the design as far as trial production, but his 1942 submachine gun was not adopted by the Soviet military. What it did get him was the attention of very influential people in the Soviet small-arms industry, who recognized his talent and ensured that he was first properly trained in weapons design and then put to work to exploit his innate genius. Kalashnikov was given the project of designing an *avtomat*, or assault rifle, in 7.62x39mm, a cartridge that had already been adopted by the Soviet military as the M1943. The Russians had been very impressed by the German *Sturmgewehr* (MP43/MP44/StuG44) designs and began working to improve them and adapt them to their own military doctrine. This was the project that was given to the young Kalashnikov.

In typical Russian fashion, Kalashnikov was only one of several designers working on the avtomat project. Besides Kalashnikov, several other noted Soviet small-arms designers, including Simonov (designer of the SKS) and Sudayev (who designed the PPs43 submachine gun) were competing for the honor of having their design accepted into service. Compared to these men, young Kalashnikov was a "Johnny come lately," and Sudayev's candidate weapon was considered at the time to have the best chance for adoption, as he had prototypes under test as early as 1944.

Testing of Kalashnikov's design did not begin until 1946, but it was so clearly superior in early evaluations that it immediately got attention at the highest levels of the Soviet military. Continued testing revealed some problems, but these were overcome, and Kalashnikov's design was type-classified in 1947. Sudayev's design was not totally rejected, however. The final AK used the magazine from Sudayev's candidate rifle. Despite its designation of AK-47, the AK did not formally enter service until 1949. This original AK-47 is referred in the West as the Type I AK-47. It has a stamped sheet-metal receiver with machined internal components held together by rivets. In service, however, this weapon unexpectedly turned out to be unreliable. The receiver reportedly developed cracks, and the rivets came loose; the weapon literally shook itself to pieces when fired extensively. Kalashnikov quickly designed a new machined receiver, which became known in the West as the Type II AK-47.

The Type II AK-47 began the legendary AK's reputation for reliability, but the rifle was extremely costly and laborious to produce. The Type II receiver began as a 2.6-kilogram (5.72-pound) block of solid steel; 120 machining operations later, it emerged as a 645-gram (1.4-pound) AK-47 receiver! The production process for the Type II AK-47 was so inefficient that it was unacceptable even to the Soviets. This inefficiency and cost led to the development of more efficient production methods, resulting in the definitive AK-47, known in the West as the Type III. The detailed differences between the varying types of AK-47 are beyond the scope of this book. If the reader is curious about the detailed history of Kalashnikov weapons and seeks more detail on the many different types, there are several excellent books on the subject.

The intermediate version of the AK assault rifle family is the AKM (Avtomat Kalashnikova Modernizirovanniyi), which marked the return to a stamped receiver. The new design was far more complex than the original AK-47 stamped receiver and required many operations to complete, but it reduced both production costs and weapon weight over that of the Type III AK-47, with no sacrifice in reliability, and was adopted as the AKM in 1959. The AKM also incorporated a muzzle compensator to offset muzzle rise on full-automatic fire, modifications to the receiver cover, and a new bayonet.

The final AK weapon prior to the AK-100 series was the AK-74, which was developed to reduce recoil and increase accuracy. The development of this weapon has already been discussed in the section on the AN-94.

General

The new generation of Kalashnikov rifles is designated AK-74M and AK-101 through AK-105. These weapons will be discussed in further detail below. The question arises, however, as to why the Russians have taken up new designations for their rifles. The designation was not arbitrary. First, all of the new generation of Kalashnikov rifles have a folding buttstock. (The short folding stock version of the AK-74, the AKS-74U, is not manufactured by Izmash, but by Tula KBP, although it was designed by the Kalashnikov Bureau at Izmash.) The folding stock distinguishes the AK-74M from the original AK-74, hence the designation. The other members of the new AK generation, the AK-100s, are based on the AK-74, but in different calibers. According to Izmash representatives, the designation for the new rifles was derived from the AK (Avtomat Kalashnikova) designation for Kalashnikov weapons combined with the number 100, an old manufacturing code for the Izmash Armory. Models of the new generation weapons were designated by successive numbers. The AK-74M, AK-101, and AK-103 all have standard-length 415mm (16.3 inch) barrels, while the AK-102, AK-104, and AK-105 all are folding-stock carbines with 314mm (12.3 inch) barrels. There is no weapon actually designated AK-100. The Russians refer to all of these weapons as submachine guns, but this is a misnomer, because the term *sub-*

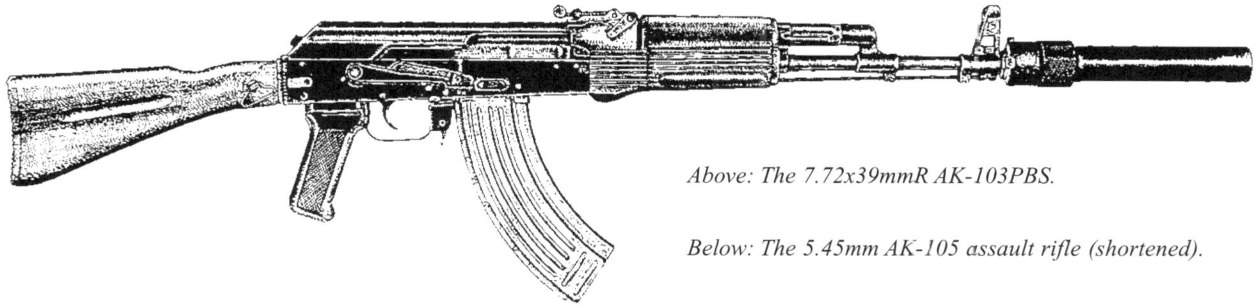

Above: The 7.72x39mmR AK-103PBS.

Below: The 5.45mm AK-105 assault rifle (shortened).

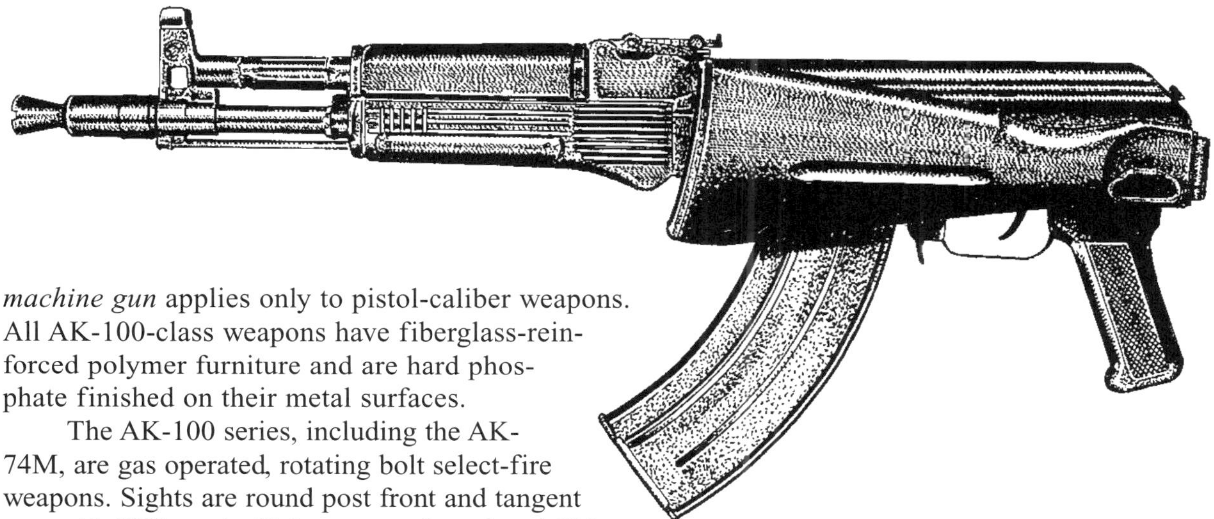

machine gun applies only to pistol-caliber weapons. All AK-100-class weapons have fiberglass-reinforced polymer furniture and are hard phosphate finished on their metal surfaces.

The AK-100 series, including the AK-74M, are gas operated, rotating bolt select-fire weapons. Sights are round post front and tangent rear with "U" notch. Sights are graduated to 1,000 meters, but this is hugely optimistic for weapons of this class, whose actual effective range in the hands of typical soldiers is no more than 300 to 500 meters. There is a standard mount on the left side of the receiver for installation of optical sights or night-vision devices.

These weapons represent what is probably the ultimate development of the superb Kalashnikov design. No matter how good a firearms design, there is only so much that can be done to effectively upgrade it, and there is little remaining that may be done to further improve the basic Kalashnikov design, which is truly the most successful small arm in military history both in terms of longevity and numbers produced. With the incorporation of modern materials and production processes into the design, the AK-100-series weapons have been upgraded to thoroughly modern standards, which will ensure their presence in the world's military forces for the foreseeable future.

The final Kalashnikov weapon that I will discuss is the AKS-74U/UB. Although this specialized version of the AK-74 has been in service for some years, there have been very few data released about it. The PBS-1 suppressor of the AKS-74U is conventional, with baffles and a rubber plug that must be replaced after only a few rounds are fired through it. An unusual feature of the AKS-74U is the fact that it incorporates the BS-1 suppressed grenade launcher (see Chapter 5)

5.45x39mm ammunition for the AKS-74U is subsonic and fires a special heavy bullet of greater sectional density than standard 5.45x39mm rounds. This special cartridge is discussed in the ammunition section of this book.

A second version of the AKS-74U is the AKS-74UB, which, according to Russian press articles, fires a cartridge similar to that of the PSS-1 silent pistol (see Chapter 2). This weapon is virtually unknown except for a few references to it in Russian firearms books and press articles. To the best of my knowledge, the AKS-74UB has never been illustrated or publicly shown. It can be assumed that the AKS-74UB has no external suppressor, because the cartridge obviates the need for one. It can also be assumed that the internal mechanism of the AKS-74UB has been completely redesigned because gas-trap cartridges cannot possibly function in a gas-operated rifle. This being the case, the AKS-74UB is probably blowback operated.

AK-100 SERIES SPECIFICATIONS

	AK-74M/101/103	AK-102/104/105
Caliber	5.45x39/5.56x45/ 7.62x39mm	5.56x45/7.62x39/ 5.45x39mm
Operation	Gas, Select Fire	Gas, Select Fire
Weight	7.48 lbs. (3.4 kg.)	6.8 lbs. (3 kg.)
Length, Extended	37.1 in. (943mm)	32.4 in. (824mm)
Folded	27.5 in. (700mm)	23 in. (586mm)
Barrel Length	16.3 in. (415mm)	12.3 in. (314mm)
Effective Range	800 m.	800 m.
Cyclic Rate	600 RPM	600 RPM
Magazine Capacity	30 Rounds	30 Rounds

OTs-14 *GROZA* (THUNDERSTORM)

During the Chechnya unpleasantness in Russia, several news photos showed Russian Ministry of Internal Affairs troops armed with what appeared to be bullpup variants of a Kalashnikov weapon, but details of the design were not forthcoming. What the compact weapon turned out to be was the OTs-14 *Groza* small-arms system. The OTs-14 is truly a "weapon system" because it can be arranged by the operator in any one of four configurations. The weapon became a favorite of MVD troops long before they employed it in Chechnya, but it was in that conflict that the rifle first achieved notoriety, and thereafter it was adopted by the military for special operations use. The OTs-14 will be with the Russian military and Interior Ministry for quite some time for reasons that will become apparent later in this section.

The OTs-14 began life as an idea of Valery Telesh, who is better known for his design of the GP-25 and GP-30 underbarrel grenade launchers. Telesh noticed the effectiveness of the U.S. M203 but felt that the weapon was hampered by many drawbacks for combat use, not the least of which was that the rifle-grenade launcher combination was overly heavy and cumbersome. Telesh believed that the addition of the M203 destroyed the balance of the weapon and made it too hard to use. He therefore set out to design an integrated system that would incorporate all the desirable features of a close-combat arm into one manageable package, using the basic Kalashnikov rifle as a starting point. Telesh and a gentleman named Yuri V. Lebedev began work on the project in December 1992; prototypes were ready in less than a year, and the OTs-14 was ready for production early in 1994—a remarkable record for development of a small-arms system such as the OTs-14.

The original weapon was conceived as being able to use any one of four cartridges: the 5.45x39mm, the 5.56x45mm, the 7.62x39mm, or the 9x39mm. The latter cartridge is used in a number of Russian special weapons and is fully described in the ammunition chapter of this book.

Telesh and the Tula Armory thought they had a winner, but nobody noticed, and the weapon seemed destined to end up like so many other Russian small arms designs—relegated to the files and museum of the armory where it originated. Apparently it was at the last minute that the OTs-14 came to the attention of the Russian MVD, which had a requirement for a compact weapon for urban combat, based on actual demands from operational units. The OTs-14 was tested and found to be exactly what the MVD troops were seeking, and it was adopted by them for use by rapid-reaction forces and special-purpose militia (police) units. It was in the hands of these special troops that the 9x39mm OTs-14 made its public debut in Chechnya.

The OTs-14 came to the attention not only of the news media, but also of the Russian Defense Ministry, which also had a requirement for such a weapon. After testing, the Defense Ministry adopted the OTs-14 in 7.62x39mm for SPETsNAZ forces use. The primary difference between the MVD and military versions of the OTs-14 is the caliber: the military version was adopted in 7.62x39mm rather than 9x39mm. (I will discuss later why the OTs-14 weapons are chambered for these calibers rather than others.)

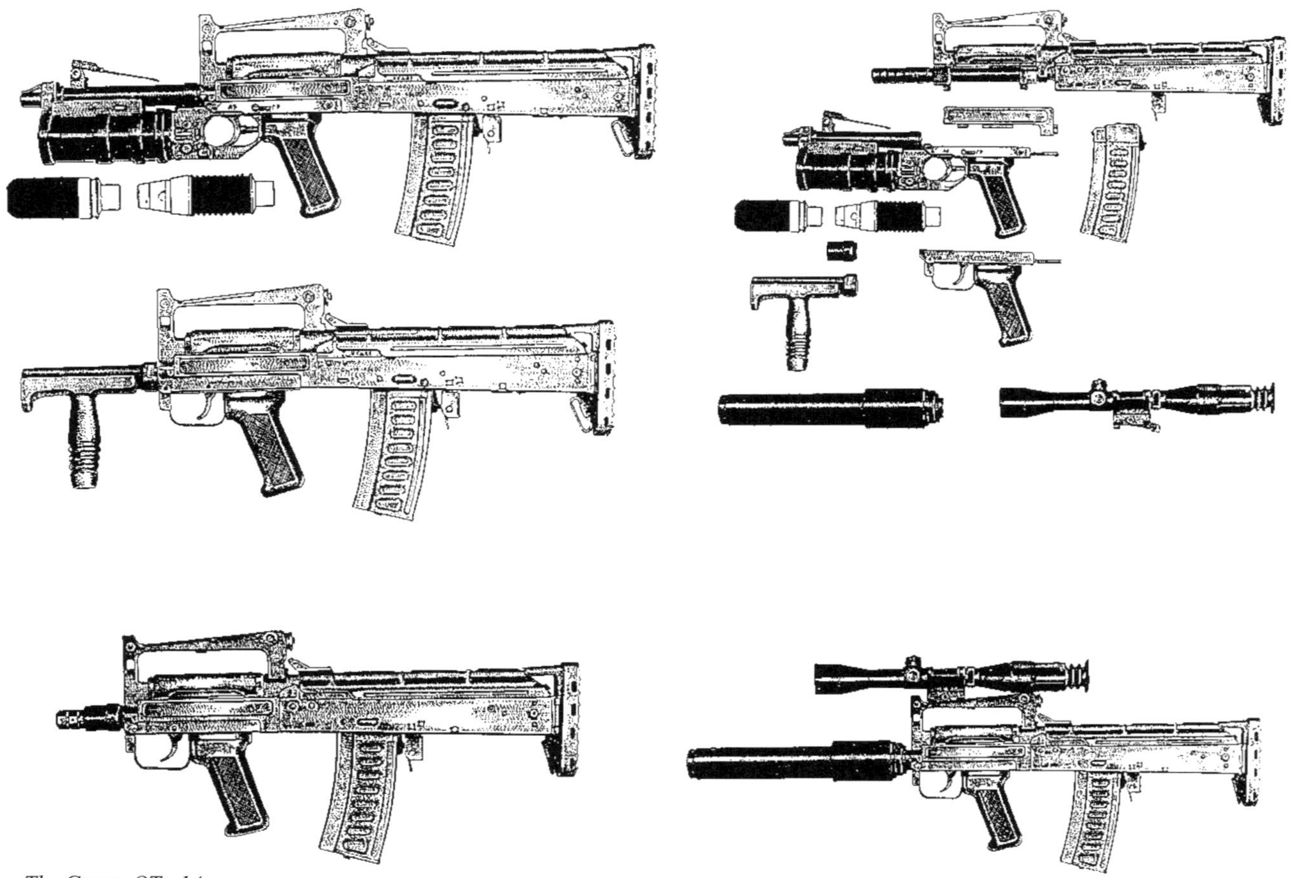

The Groza *OTs-14 family of weapons.*
Top Left: Groza *OTs-14-4A basic version.*
Middle Left: Groza *OTs-14-4A-01 special assault rifle.*
Bottom Left: Groza *OTs-14-4A compact version.*
Top Right: Groza *OTs-14-4A special weapons system component parts.*
Bottom Right: Groza *OTs-14-4A-03 special-mission version.*

The OTs-14 is issued in an aluminum carrying and storage case and is equipped with virtually all of the necessary equipment to adapt it for any imaginable mission. There are two different grip and trigger assemblies, one for use with the GP-25/30 grenade launcher and another for use when the grenade launcher is detached. When the grenade launcher is attached, the single OTs-14 trigger fires both it and the rifle. A selector on the left side of the grip allows the operator to select either the rifle or grenade barrel. When the grenade launcher is detached, a vertical grip is fitted to the barrel. A suppressor is standard issue, as is a telescopic sight which mounts directly onto the carrying handle. There is also a quick-change barrel for use with the suppressor or when maximum compactness is required. Interestingly, the OTs-14 is given a different designation for each of its four configurations. The full designation of the basic version of the OTs-14 with grenade launcher in place is OTs-14-4A. When the grenade launcher is removed, the designation changes to OTs-14-4A-01. Changing to the short-barrel version alters the designation to OTs-14-4A-02, and adding the suppressor gives the designation OTs-14-4A-03. Specific designations for each configuration is probably so that operators can be quickly informed as to which configuration to use for a specific

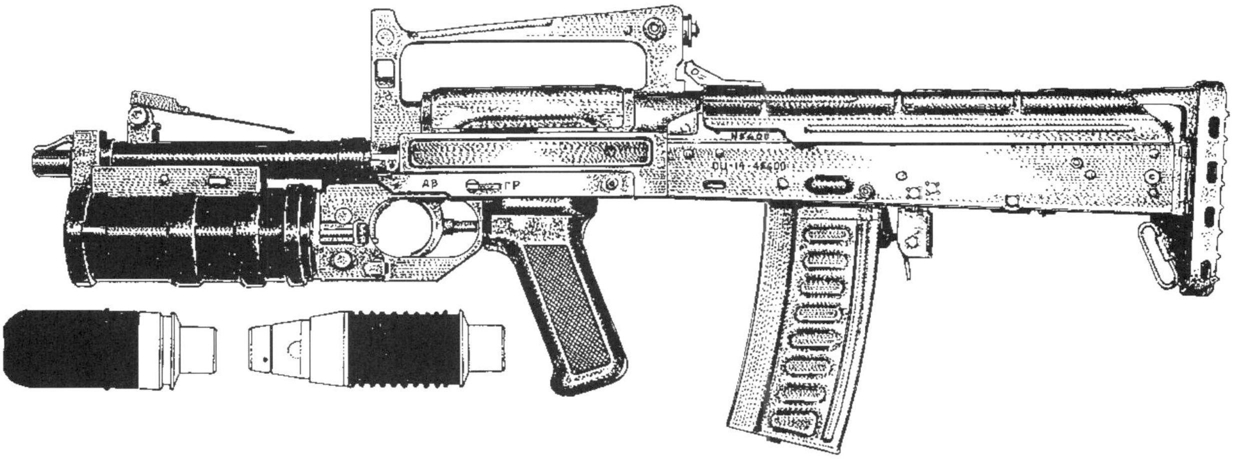

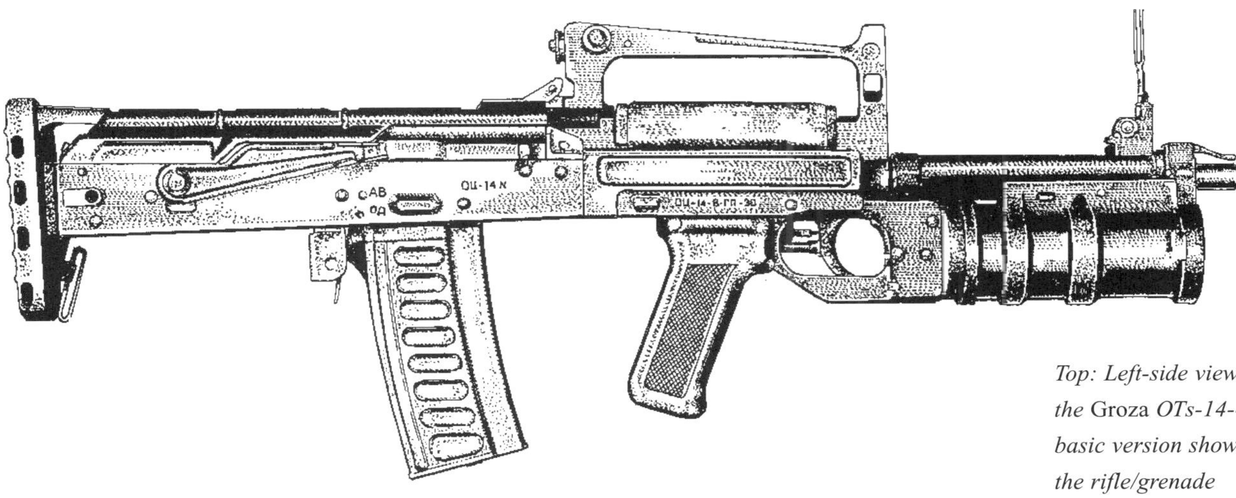

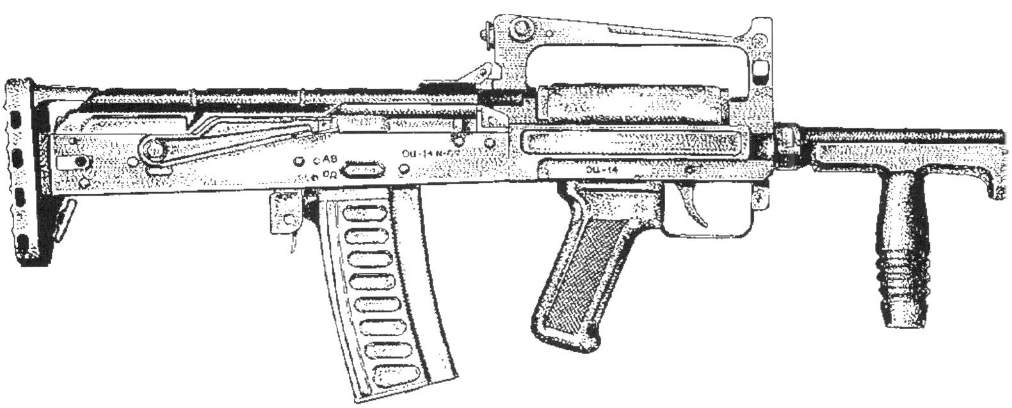

Top: Left-side view of the Groza *OTs-14-4A basic version showing the rifle/grenade selector lever above the trigger guard, baton round, and VOG-25P.*

Middle: Right-side view of the Groza *OTs-14-4A basic version.*

Left: Short-barreled version of the Groza.

The 9x39mm OTs-14-4A-01.

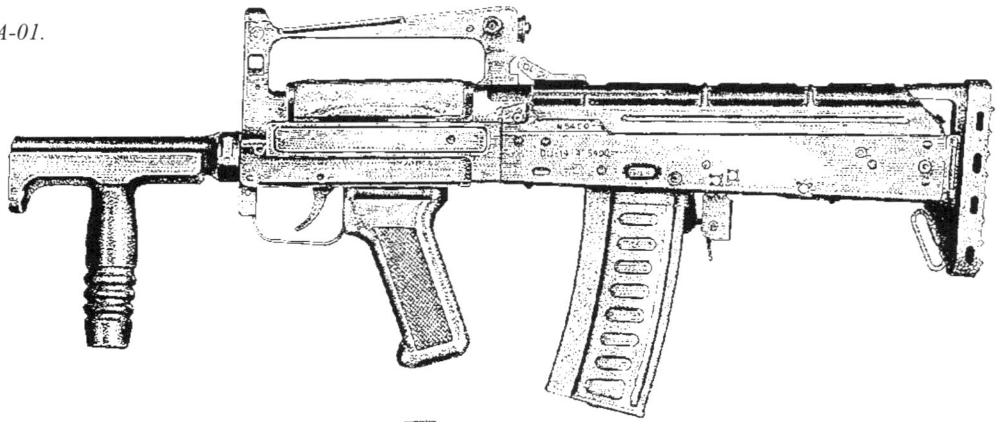

The Groza *OTs-14-4A-02 compact version.*

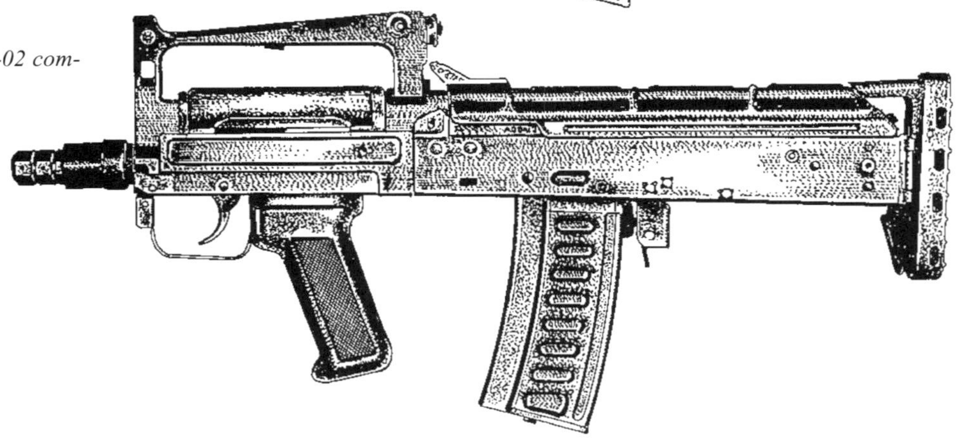

Left-side view of the Groza *OTs-14-4A-03 special-mission version.*

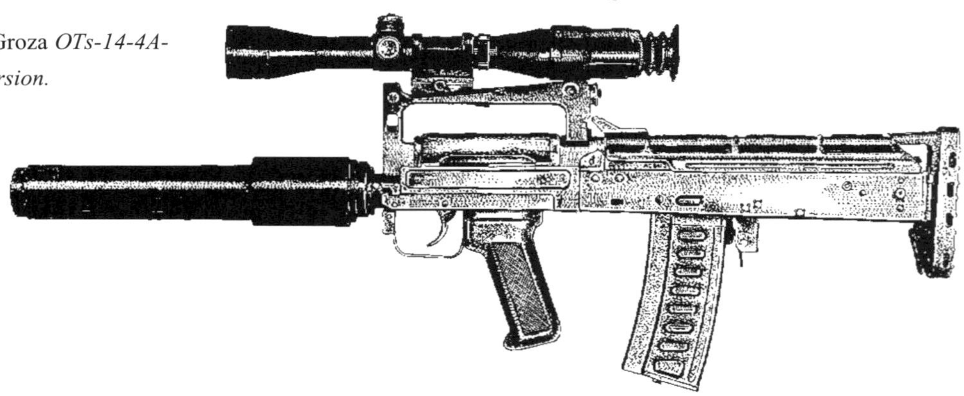

Right-side view of the Groza *OTs-14-04-4A-03 special-mission version.*

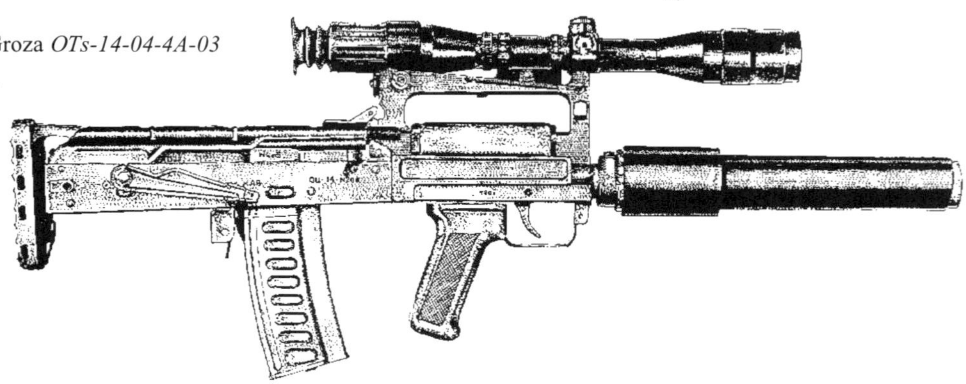

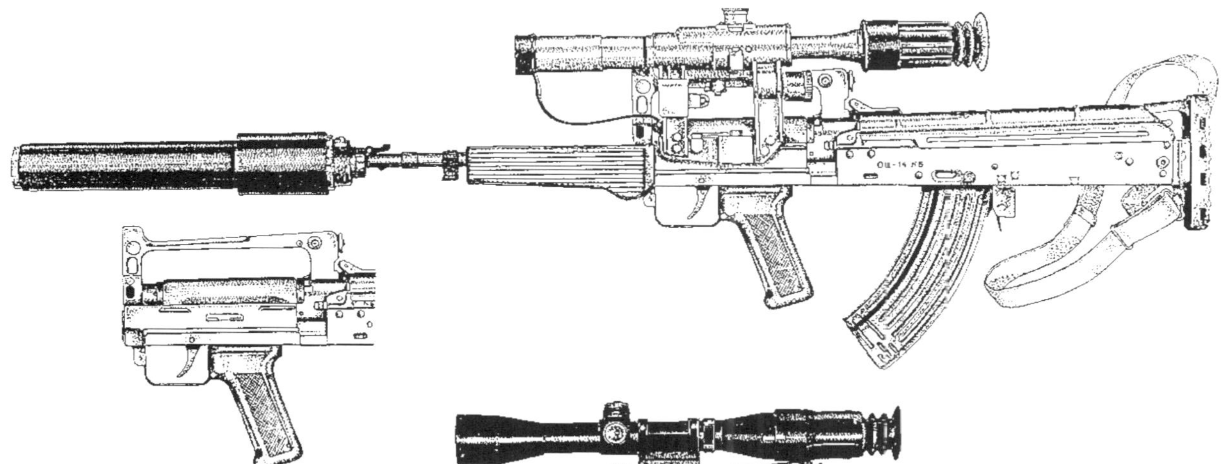

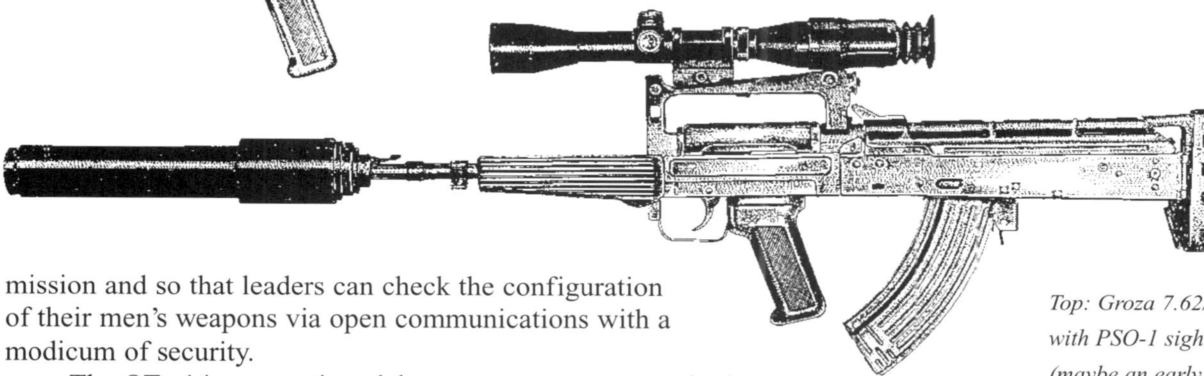

Top: Groza 7.62/40 with PSO-1 sight (maybe an early type or prototype). Inset shows the PSO-1 brackets in the receiver.

Bottom: Groza with commercial sights (notice that the bottom weapon has no fittings for PSO-1).

mission and so that leaders can check the configuration of their men's weapons via open communications with a modicum of security.

The OTs-14, as mentioned, has a suppressor as standard equipment. Suppressors are used by special-operations forces on raids so that they can communicate with each other, reduce sound signature, and keep muzzle flash to a minimum. They may be forced to rapidly install or remove their suppressors to meet changing mission needs. The OTs-14 fulfills this requirement by incorporating an interrupted-thread "snap-on" interface for the suppressor. The use of a suppressor also mandates the use of ammunition that does not depend upon high velocity for its effectiveness. High-velocity ammunition invariably leaves the muzzle of the weapon from which it is fired at supersonic velocity and as a result, there is a "crack" at the muzzle as the weapon is fired and the bullet breaks the sound barrier. Using a suppressor with such a round would be only partly effective and would still give a noise signature for adversaries to shoot at. The use of subsonic ammunition eliminates the sonic "crack," but small, light bullets designed to achieve terminal effects via high velocity are virtually useless at subsonic velocities. The Russians, therefore, developed the 9x39mm cartridge for use in special operations and suppressed weapons. There is also a special subsonic round in 7.62x39mm, and it costs less to produce than the 9x39mm round and standard 7.62x39mm can be used for training, hence its adoption for the OTs-14 by the Russian military. As previously mentioned, both of these rounds are covered in detail in the ammunition chapter of this book.

The OTs-14 is not for general issue, but is intended for special operations forces. According to an article in a Russian periodical, it is "a weapon for high class professionals." Whatever its purpose, the OTs-14 is a light, compact, versatile weapon of ingenious design. It will certainly be a part of the SPETsNAZ and MVD inventory for years to come.

OTs-14 *GROZA* SPECIFICATIONS

	Military Version	**MVD Version**
Caliber	7.62x39mm	9x39mm
Operation	Gas, Select Fire	Gas, Select Fire
Weight	6.82 lbs. (3.1 kg.)	5.94 lbs. (2.7 kg.)
Length	27.5 in. (700mm)	22.8 in. (580mm)
Barrel Length	App. 18 in. (457mm)	App. 12 in. (304mm)
Effective Range	600 m.	400 m.
Cyclic Rate	750 RPM	700 RPM
Magazine Capacity	30 Rounds	20 Rounds
Sights	Front, Blade; Rear, Notch	Front, Blade; Rear, Notch

SVD, SVDS, AND SVU DRAGUNOV SYSTEMS

The SVD Dragunov is well known in the West, so we will not go into great detail about its development and design. The SVD was adopted by the Soviet military in 1963, and although it was not specifically designed by Kalashnikov, it is obvious that Dragunov started with the AK as a baseline and went from there to develop a reliable semiautomatic "sniper" rifle in the Russian/Soviet tradition, which is markedly different from that of Western armies.

Russian snipers are really *marksmen* in the Western meaning of the word, and their abilities are far removed from those of Western snipers. The typical engagement range of Soviet/Russian snipers is no more than 500 meters, whereas Western snipers regularly engage targets out to 1,000 meters. When the SVD was first examined by Western technical intelligence personnel, they were surprised at the rifle's lack of precision accuracy. It was only when they noted the Soviet tactical employment of the rifle that they realized that it was more than satisfactory from the Russian tactical standpoint. The legend about the phenomenal accuracy of the SVD, to the effect that it is more accurate than Western bolt-action sniper rifles, is patently false. In fact, other semiautomatic precision rifles such as the Armalite AR-10 (T) and Heckler & Koch's PSG-1 achieve accuracy far superior to that of the SVD.

Curiously, Izmash experimented with a fully automatic version of the SVD, which never reached production. Izmash's failure to produce this version of the SVD is not surprising; a fully automatic sniper rifle is something of an oxymoron, although some machine guns achieve excellent levels of accuracy when fired on semiautomatic.

The 7.62mm SVDS with PSO-1 sight. Points to note are plastic pistol grip, plastic handguard, new short flash hider, and rubber cheek pad on folding stock. The rifle is not designed to fire with the stock folded. Both the cocking handle and trigger are obstructed by the stock.

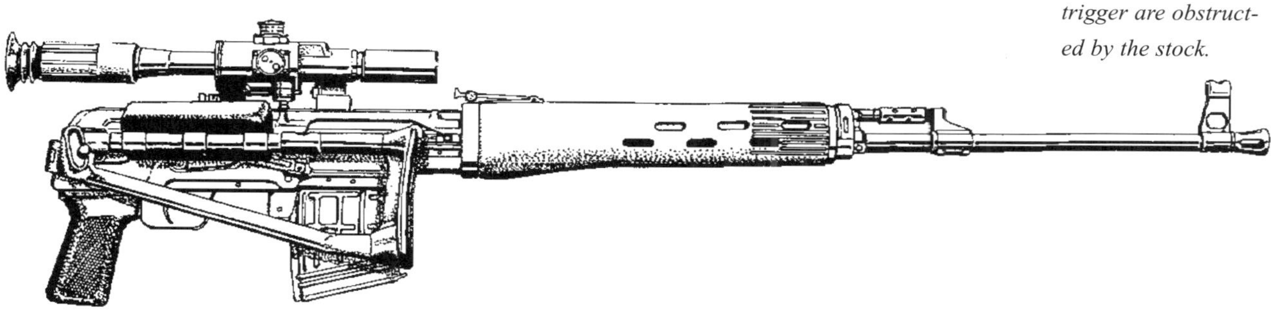

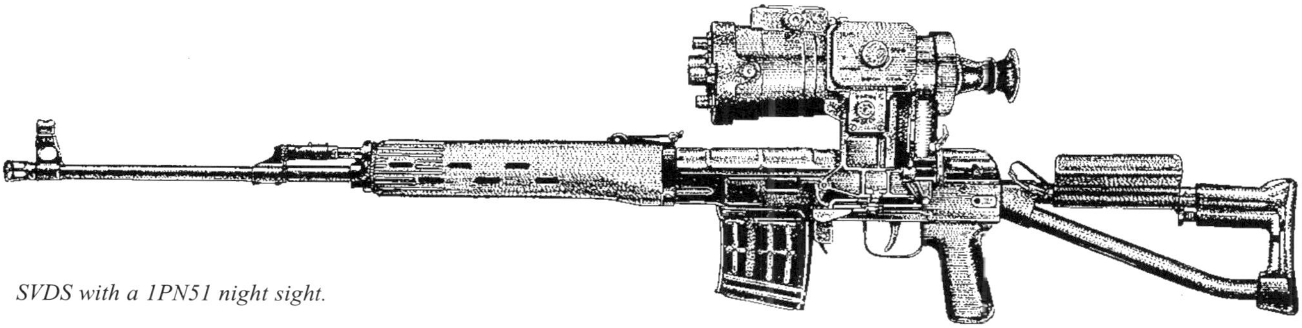

SVDS with a 1PN51 night sight.

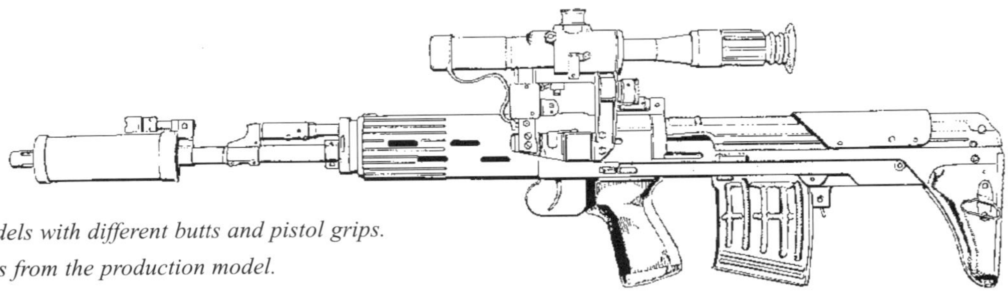

Preproduction models with different butts and pistol grips. The foregrip differs from the production model.

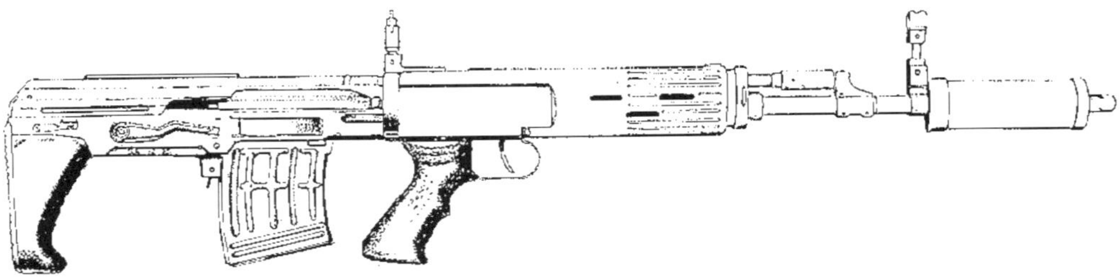

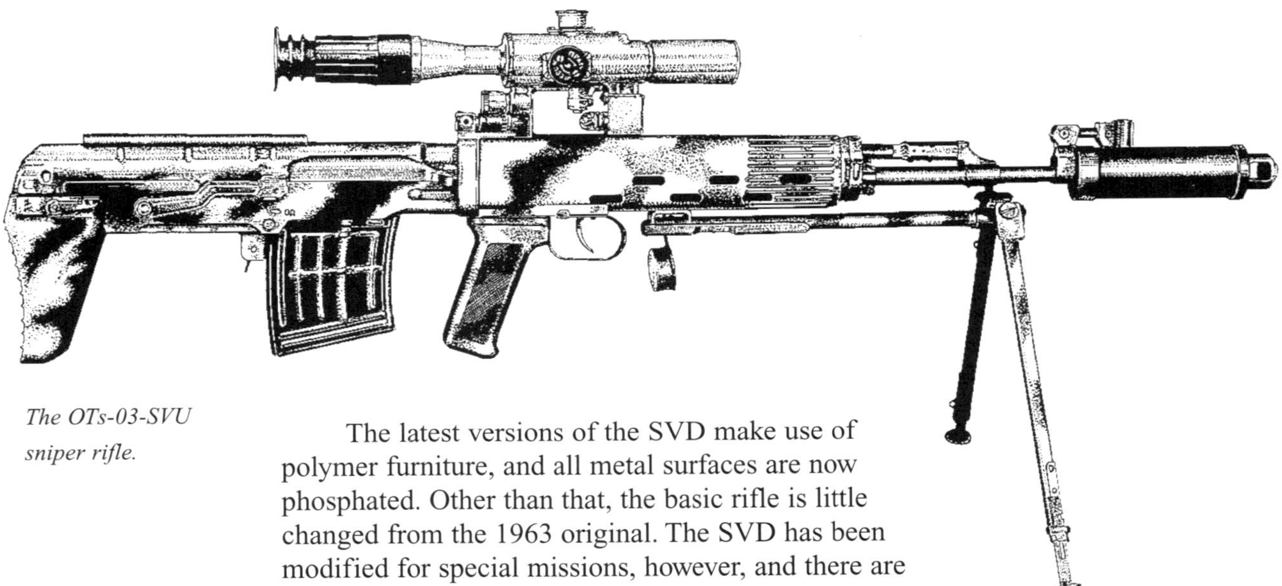

The OTs-03-SVU sniper rifle.

The latest versions of the SVD make use of polymer furniture, and all metal surfaces are now phosphated. Other than that, the basic rifle is little changed from the 1963 original. The SVD has been modified for special missions, however, and there are now three different versions of the rifle: the basic SVD, the SVDS, and the OTs-03 SVU. The latter rifle is manufactured by Tula KBP, rather than Izmash. The SVD, as mentioned, is well known and has even been sold commercially in the West, so it will not be further discussed.

The SVDS is less well known than the SVD, but it has been offered for military sale and has been reported on in the Western press. The SVDS is intended for use by airborne and motor-rifle (mechanized) troops. The SVDS fires the same 7.62x54R ammunition as the SVD, but the presence of the following features distinguish it from the basic Dragunov:

- Folding stock
- Adjustable cheekpiece
- Shorter, heavier barrel
- Different flash hider
- Pistol grip
- No bayonet lug

The folding stock of the SVDS is made of tubular steel and folds to the right. In the "carry" or "march order" position, the stock is folded, and the adjustable cheekpiece is rotated down. When the stock is open for firing, the adjustable cheekpiece is rotated "up" for using the telescopic sight and "down" if the backup open sights are to be used. The pad itself is made of fiberglass-reinforced polymer.

The new flash suppressor is interesting in that it is much smaller than the original SVD flash suppressor, which remains in use on the latest SVD rifles. The SVDS suppressor is not only shorter and conical, it has asymmetrical, teardrop-shaped slots and a wide web at the bottom to preclude muzzle rise and prevent dust from being kicked up when the rifle is fired. Data on both the SVD and SVDS are contained in the table at the end of this section.

The OTs-03 SVU is the sniper rifle counterpart of the OTs-14 discussed above and is made by Tula KBP. Oddly for a sniper rifle, the OTs-03 is manufactured in a select-fire version. As mentioned, Izmash produced a few select-fire SVD rifles in the 1980s, but they were experimental and never went into full production. The OTs-03 is intended for use by special forces of both the MVD and military, the idea being to provide the user with a rifle that combines the firepower of an assault rifle with the accuracy of a sniper rifle. This select-fire capability is not only unusual but highly questionable, given the fact that the SVD is chambered for the 7.62x54R cartridge, which probably is very difficult to control on full automatic. Moreover, the OTs-03 uses the standard 10-round SVD magazine, and, at a cyclic rate of fire of 880 RPM, 10 rounds can be fired almost instantly.

There has been very little written about this rifle in the open press. As was the OTs-14, the OTs-03 was used by MVD troops in Chechnya, and, as is the OTs-14, the OTs-03 is produced by Tula KBP. Although there have been no specific statements to the effect in the Russian press, the OTs-03 was also probably designed by Valery Telesh. It should be noted that the OTs-03 has almost invariably been observed with a suppressor attached, leading to the conclusion that this is part of the overall OTs-03 weapons system. The Russians have also probably developed subsonic 7.62x54Rmm ammunition for use in suppressed weapons.

The OTs-03 uses the tried and true PSO-1 4x telescopic sight that has been mounted on the SVD rifles since their adoption in 1963. The PSO-1 is also used on virtually every other Russian rifle that uses a telescopic sight, even the massive V-94 and OSV-96 12.7x108mm antimateriel rifles (see below), albeit in 13x rather than the usual 4x. Like the other variants of the SVD, the SVU has backup iron sights, but those of the SVU fold down when not in use and are quite different from those of either of the other rifles.

In sum, the SVU is clearly another special-purpose rifle whose use will be limited to special-operations units. Even so, the short barrel, which appears to be approximately 16 to 18 inches in length, cannot help but reduce accuracy to some extent, and there appears to be overlap between the mission profile of the SVU and the SVDS. Perhaps the Russian special forces, like their Western counterparts, can purchase almost any weapon they desire without regard for standardization regulations. Whatever the mission need that drove its development and fielding, the SVU/OTs-03 is an interesting spinoff of the Dragnuov.

SVDS/OTS-03 SVU SPECIFICATIONS

	SVDS	**OTs-03 SVU**
Caliber	7.62x54mm	7.62x54mm
Operation	Gas, Semiautomatic	Gas, Select Fire
Weight	10.3 lbs. (4.68 kg.)	12.1 lbs. (5.5 kg.)
Length, Extended	44.7 in. (1,235mm)	35.4 in. (900mm)
Folded	34.5 in. (875mm)	NA
Barrel Length	22.3 in. (565mm)	App. 20 in. (508mm)
Effective Range	1,200 m.	800 m.
Cyclic Rate	NA	NA
Magazine Capacity	10 Rounds	10 Rounds
Sights	Optical, Iron Backup	Optical, Iron Backup

AS *VAL* (RAMPART) SILENT ASSAULT RIFLE AND THE VSS VINTOREZ (THREAD CUTTER) SILENT SNIPER RIFLE

It is not clear how small-arms nicknames are derived in Russia, but they tend to be unusual, to say the least. These two rifles are clear indicators of that fact. The *Val* and *Vintorez* are essentially the same rifle with modifications for different tactical employment. Both share receivers with the MA *Vikhr* (see p. 37) compact assault rifle, also chambered for the 9x39mm cartridge. According to Russian sources, both the AS and VSS rifles have been in use by SPETsNAZ and MVD forces for some years. As stated, the rifles are, for all intents and purposes, identical save for two differences: the AS cannot be completely disassembled and covertly packed into a briefcase, and the AS has a folding metal stock similar to that of the SVDS and a pistol grip, rather than a "Dragunov-style" fixed skeleton wood stock like the VSS. Having described the design differences of the two rifles, I will treat them together for the balance of this section.

The rifles were designed at TsNIITochmash, located at Klimovsk, by a team headed by Pyotr Ivanovich Serdyukov. According to Russian sources, Serdyukov carefully studied ballistics before designing his silenced rifles and arrived at a suppressor design that reduces not only noise and gas pressure but heat as well. According to a 1994 article in *Krasnaya Zvezda (Red Star),* the official Russian military newspaper:

> The surprising success of this weapon comes from its relationship between the two main enemies of noiseless firing: pressure and heat. Gases, once the bullet has immediately cleared the muzzle, begin to immediately fan out in all possible directions; as they do, there is an intense increase in pressure in the immediate area. But here—in the first chamber of the silencer, where they disperse—the simultaneous reduction in temperature also has an immediate effect upon the pressure. Finally, the item referred to as the separator ultimately "breaks up" the wave which follows the bullet, by running it through a multi-baffle set which sends the shock wave in various different directions. The set ensures that it always stays cool. These are the principles. (*Krasnaya Zvezda*, 4 March 1994, p. 2)

Although this hardly clears up the design principles of the AS/VSS suppressor, it appears to be fairly conventional. Whatever the technology involved in its construction, the AS/VSS suppressor is effective. The author has fired both weapons, and at the time literally hundreds of rounds were put through each with no apparent degradation of suppression. The Russian hosts would not, however, allow the suppressors to be taken apart, nor would they answer any detailed questions about their construction.

The design of both rifles is conventional, apart from their suppressors. They are gas-operated, select-fire, magazine-fed compact rifles with very short ventilat-

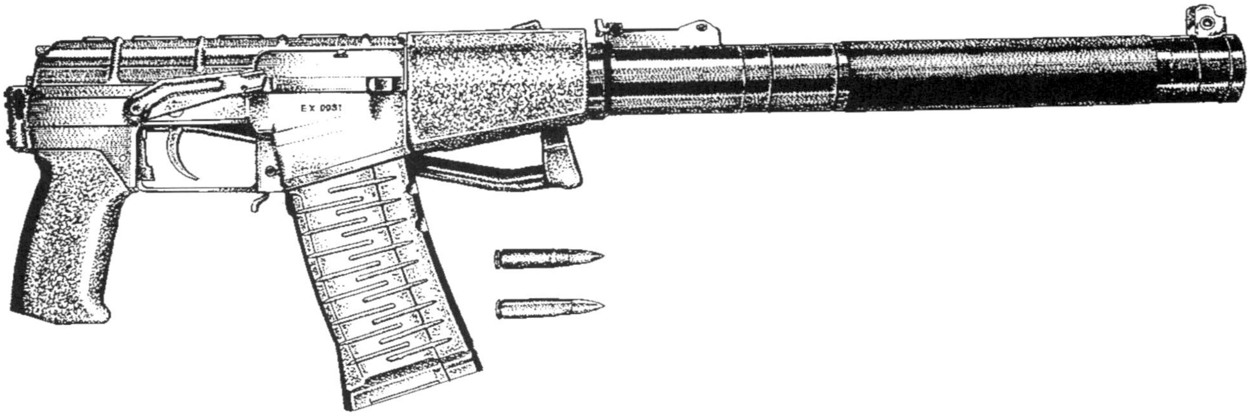

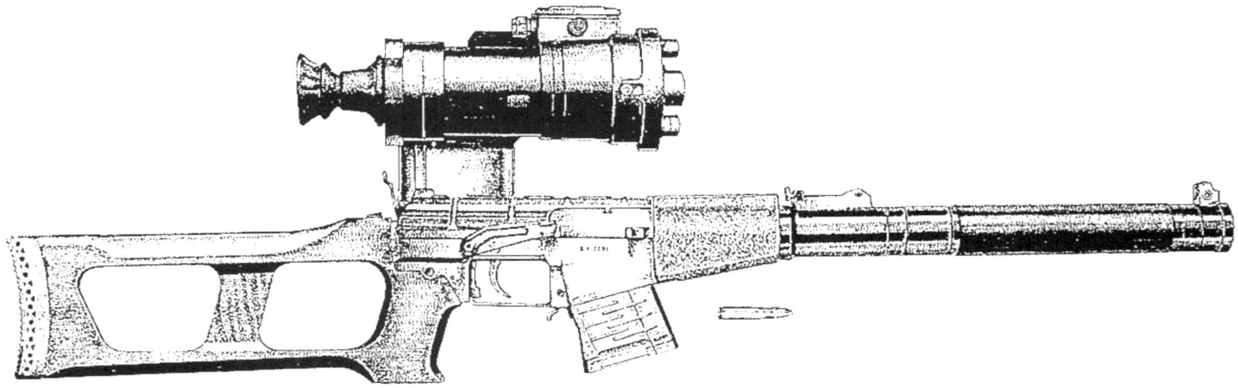

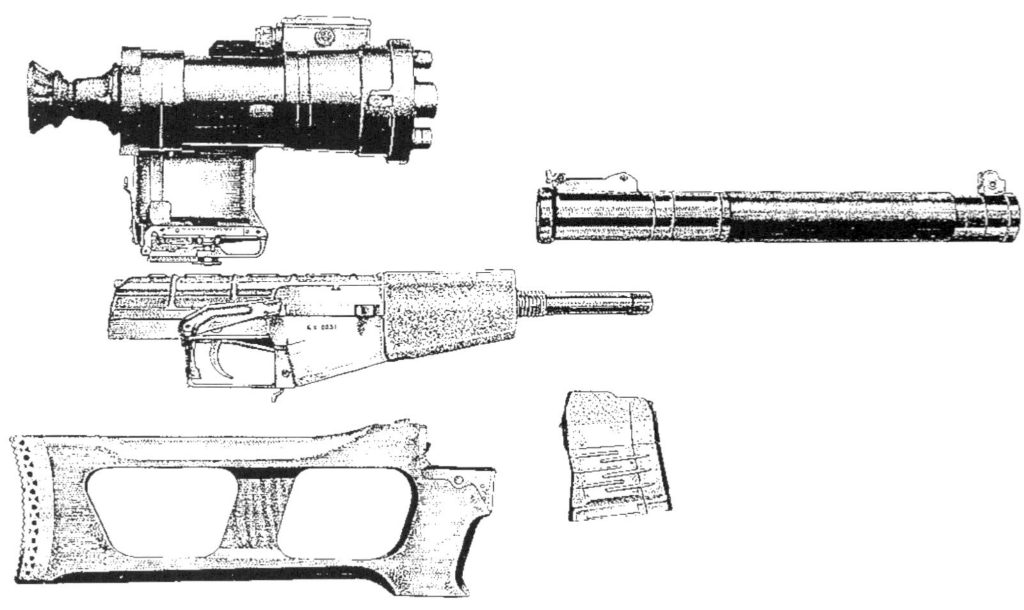

Top: AS Val *special assault rifle in 9x39mm (AP).*

Above: VSS Vintorez *in 9x39mm with 1LH51 second-generation night sight.*

Right: VSS rifle taken down for discreet transport.

ed barrels covered by an integral suppressor. From their design it is clear that they are not intended to be fired unsuppressed. Both are obviously Kalashnikov derived, but with some significant differences. The receiver cover, for example, is nondetachable. When the release on the rear of the cover is pressed, it hinges up at the front to reveal the weapon's operating mechanism, which is similar to that of the Kalashnikov, although the gas system is quite different and the weapons are striker fired rather than hammer fired like the AK-series weapons. The weapons apparently will not function without the suppressors in place because they were designed with the suppressors as an integral part of the operating system, which does not generate sufficient gas pressure for system operation without the suppressor.

As mentioned previously, the term *sniper* has different connotations for the Russian military than it does for its Western counterparts. With this in mind, the 400-meter effective range of the VSS should come as no surprise. The author has fired this rifle at ranges of up to 200 meters, and it is accurate to approximately 2 minutes of angle at that range. There is every reason to believe that it would hold similar accuracy for another 200 meters, given its ammunition. This is hardly within the purview of Western sniper weapons, but well within the standards for Russian snipers whose targets (and the ranges at which the snipers engage them) are quite different from those common in the West.

The VSS is issued as a part of the VSK sniper system, different from the VSK-94 described below, which comes in its own aluminum carrying case and includes a disassembled VSS rifle, a PSO-1 4x day optical sight (albeit with a reticle different from that of the PSO-1 issued for use with the SVD), a 1PN75 night-vision optic, spare magazines, and operator's tool kit. Magazines are polymer and are provided in both 10- and 20-round versions. The 10-round versions are intended for use with the VSS sniper rifle, and the 20-round magazine is intended for the AS, although the magazines are interchangeable between both rifles.

AS AND VSS SPECIFICATIONS

	AS	VSS
Caliber	9x39mm	9x39mm
Operation	Gas, Select Fire	Gas, Select Fire
Weight	5.5 lbs. (2.5 kg.)	5.7 lbs. (2.6 kg.)
Length, Extended	34.5 in. (878mm)	35.1 in. (894mm)
Folded	24.2 in. (615mm)	NA
Barrel Length	App. 12 in. (304mm)	App. 12 in. (304mm)
Effective Range	400 m.	400 m.
Cyclic Rate	App. 700 RPM	App. 880 RPM
Magazine Capacity	20 Rounds	20 Rounds
Sights	Front, Blade; Rear, Notch	Optical, Iron Backup

VSK-94 SNIPER RIFLE

The Tula Arsenal VSK-94 sniper rifle is a development of the A-91 (see below) compact assault rifle in much the same way that the VSS was derived from the MA. Why the Russians would market two weapons with essentially identical capabilities is something of a mystery unless one considers the Russian system of small-arms development.

The Russians, and the Soviets before them, typically give a military requirement to several design teams, which then compete against each other to produce the final product. This practice dates back to the Great Patriotic War (World War II). Readers will recall that Kalashnikov competed against other designers when he designed the AK-47 and also that the latest Russian assault rifle, the AN-94, was the result of a design competition. Given this background, we can surmise why there are so many virtually identical designs now being marketed by Russian small-arms manufacturers. There can be only one "winner" in any competition, and with the economic disaster that is currently plaguing Russia, the "losers" cannot simply shrug their shoulders, walk back to the factory, and go on to an alternative project. In the old days of the Soviet Union, the government meted out projects to keep each production facility viable. But in today's Russia there are very few "next projects"; the government no longer subsidizes arms manufacturers, and if one loses in a government weapons competition, there is no civilian market for the design. Manufacturers must thus seek recourse in the international arms market.

Alternatively, the Russian military and police may be purchasing both the VSK-94 and VSS/VSK systems, but this is not likely. Nonetheless, both Tula KBP and TsNIITochmash claim that their nearly identical rifles are in use by Russian Federation MVD troops.

The VSK-94 is a conventional design and differs in capability from the VSS mentioned above primarily in that the VSK-94 is intended to be fired without its suppressor in place and, unlike the TsNIITochmash weapons, will function with or without the suppressor. Otherwise, the VSK-94 is a gas-operated, magazine-fed, select-fire, compact rifle and is unusual only in its 9x39mm ammunition. The VSK suppressor appears to be similar in design to the VSS, but the VSS suppressor is integral to the design and not an add-on, as is the case with the VSK suppressors. The VSK-94 uses a variant of the PSO-1 day optic designated PSO-1-1. It is probably the same as that used by the VSS, presumably with the same reticle, because the VSK-94 fires the same 9x39mm cartridge as the VSS and has the same capabilities. Since Russian small arms use standard mounting hardware, the VSK-94 will accept any standard Russian optical device, including the PKN-032 night-vision optic and the PKS-07 collimating sight. The latter appears very much like the Aimpoint and similar 1x optics. The fixed wooden stock of early VSK-94 rifles appears to be virtually identical to that of the VSS, but latest versions of the VSK-94 have a "skeleton" polymer stock.

The gas operating system of the VSK-94 is derived from the Kalashnikov, but like the VSS, there are differences. The VSK-94 system more closely resembles the original Kalashnikov than the VSS. Gas is bled off near the muzzle through a

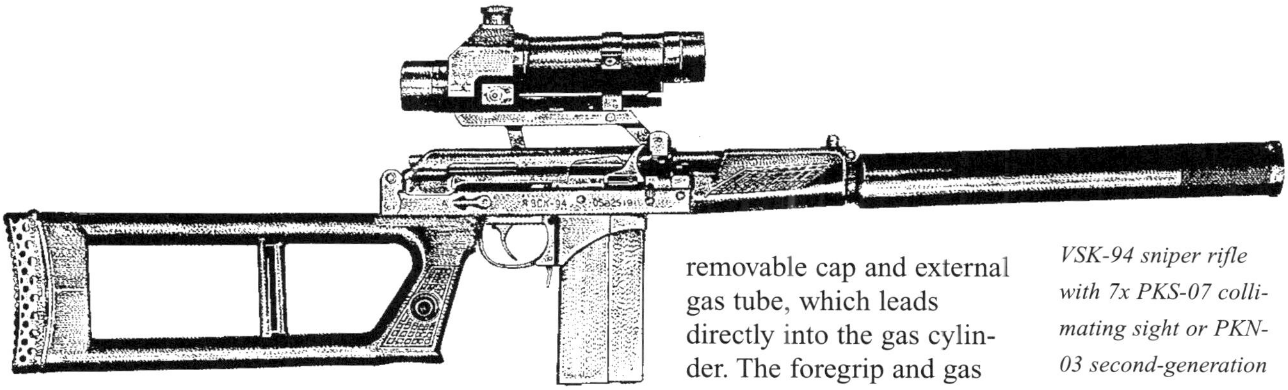

VSK-94 sniper rifle with 7x PKS-07 collimating sight or PKN-03 second-generation night sight.

removable cap and external gas tube, which leads directly into the gas cylinder. The foregrip and gas tube assembly are retained by a knurled threaded ring, which is visible in the illustrations at the front of the foregrip. The bolt cover of the VSK-94 receiver is completely removable and the bolt and carrier are shorter, while being very similar to the Kalashnikov design, although the recoil spring is retained with the bolt cover, rather than being retained with the release latch as with the Kalashnikov.

Like the VSS, the effective range of the VSK-94 is claimed to be 400 meters, and given the configuration of the rifle and its ammunition, this is probably an accurate statement.

VSK-94 SPECIFICATIONS

Caliber	9x39mm
Operation	Gas, Select Fire
Weight	5.9 lbs. (2.7 kg.)
Length	35.4 in. (900mm)
Barrel Length	App. 12 in. (304mm)
Effective Range	400 m.
Cyclic Rate	700–900 RPM
Magazine Capacity	20 Rounds
Sights	Optical, PSO-1, PKN-032 Night Vision, PKS-07 Collimating (Red Dot)

A-91 COMPACT ASSAULT RIFLE

The A-91 is actually part of a family of compact assault rifles, which was designed to an Interior Ministry requirement for a small, select-fire assault rifle/carbine that could be used for both personal defense and close-quarters battle (CQB). According to Russian sources, compact rifles such as the A-91 are necessary because the standard AK-74 rifles are both too lethal and have too great a range for urban combat against criminals. Further, the chances of ricochets and overpenetration is a problem with the standard military rifle. Pistol-caliber 9mm cartridges, on the other hand, are not sufficiently lethal. They will not penetrate body armor or car bodies. This problem resulted in the development of the 9x39mm cartridge by TsNIITochmash, which is covered in the ammunition section of this book. Briefly, the 9x39mm cartridge fires a bullet of a very high sectional density at subsonic velocity, resulting in excellent terminal effects against both personnel and material. The A-91 was originally developed by the Tula KBP to chamber this new 9x39mm cartridge and in that caliber is designated the 9A-91. Other versions of the A-91 are likewise designated according to their caliber—namely, 5.45A-91 (5.45x39mm); 5.56A-91 (5.56x45mm); and 7.62A-91 (7.62x39mm). Other than caliber and ballistics, all versions of the A-91 are identical, save for modifications necessary to accommodate the various cartridges. It is clear that economics drive the various chamberings of the A-91; Tula is advertising the A-91 on the world arms market, stressing the versatility of the weapon in its various chamberings. As of the time of this writing (early 1998), there have been no known international sales of any version of the A-91. The 9A-91 has been widely demonstrated to Western military attachés and the press at Russian military shows, and Tula KBP states that it is in service with Russian MVD troops.

The A-91 is a gas-operated carbine derived from the basic Kalashnikov design. This is not surprising; the very similar AKS-74U is manufactured by Tula, although it was designed by the Kalashnikov Bureau at Izmash. The A-91's selector switch is ambidextrous, located just above the trigger so that it can be manipulated by the user's thumb without changing grip. The A-91 is generally made of sheet-metal stampings and plastic, thereby simplifying manufacturing.

The A-91 was designed to accommodate a variety of accessories and optics, including the ubiquitous PSO-1

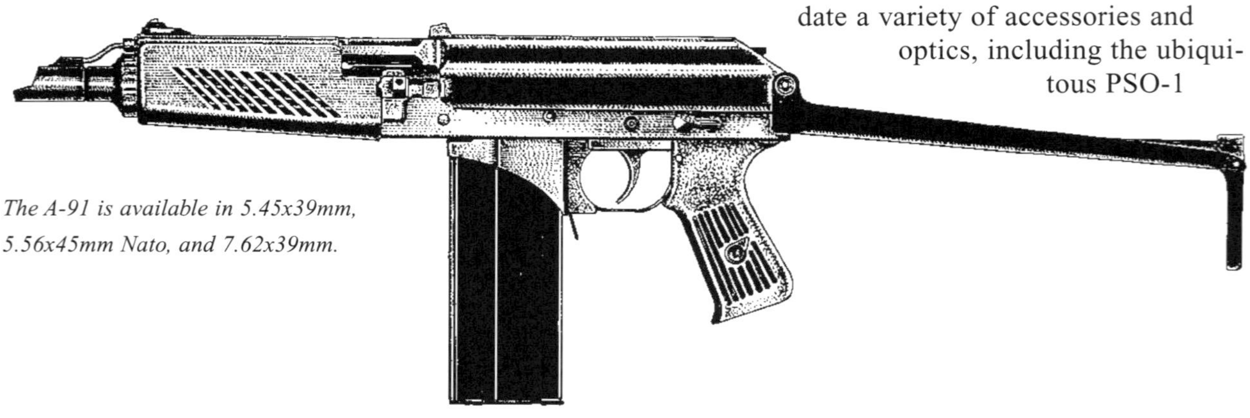

The A-91 is available in 5.45x39mm, 5.56x45mm Nato, and 7.62x39mm.

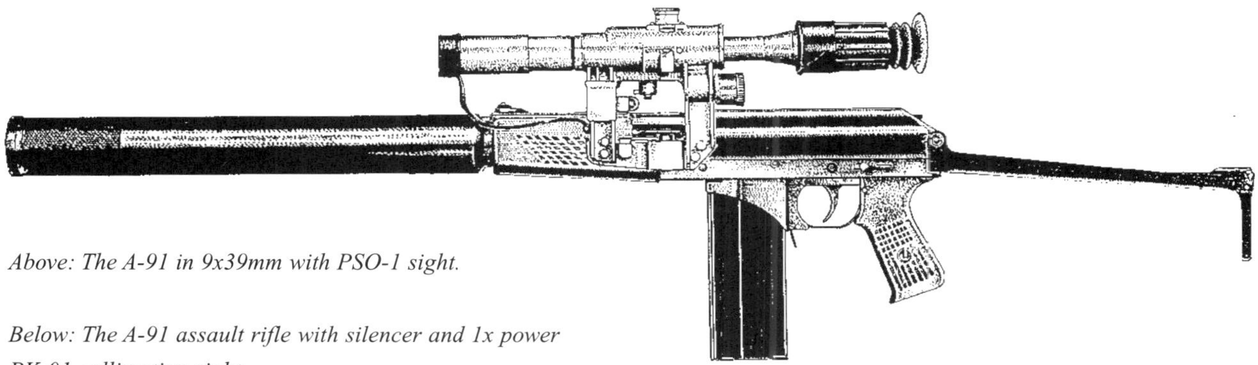

Above: The A-91 in 9x39mm with PSO-1 sight.

Below: The A-91 assault rifle with silencer and 1x power PK-01 collimating sight.

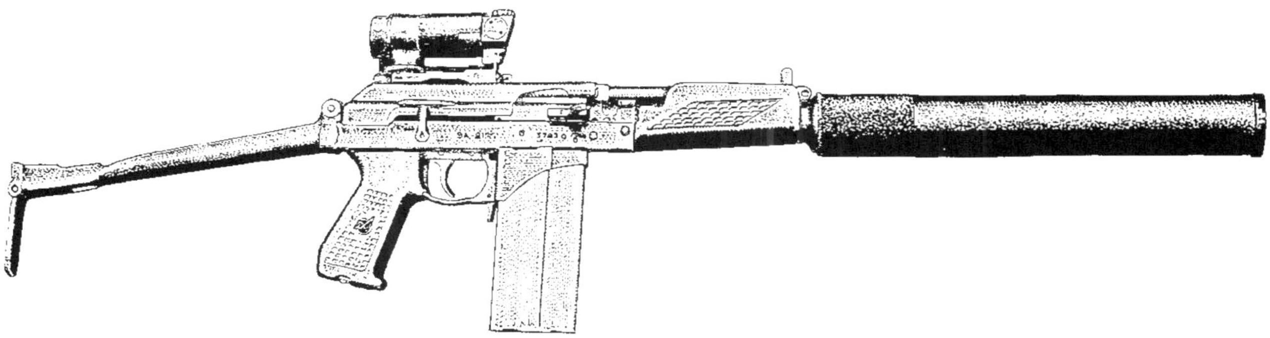

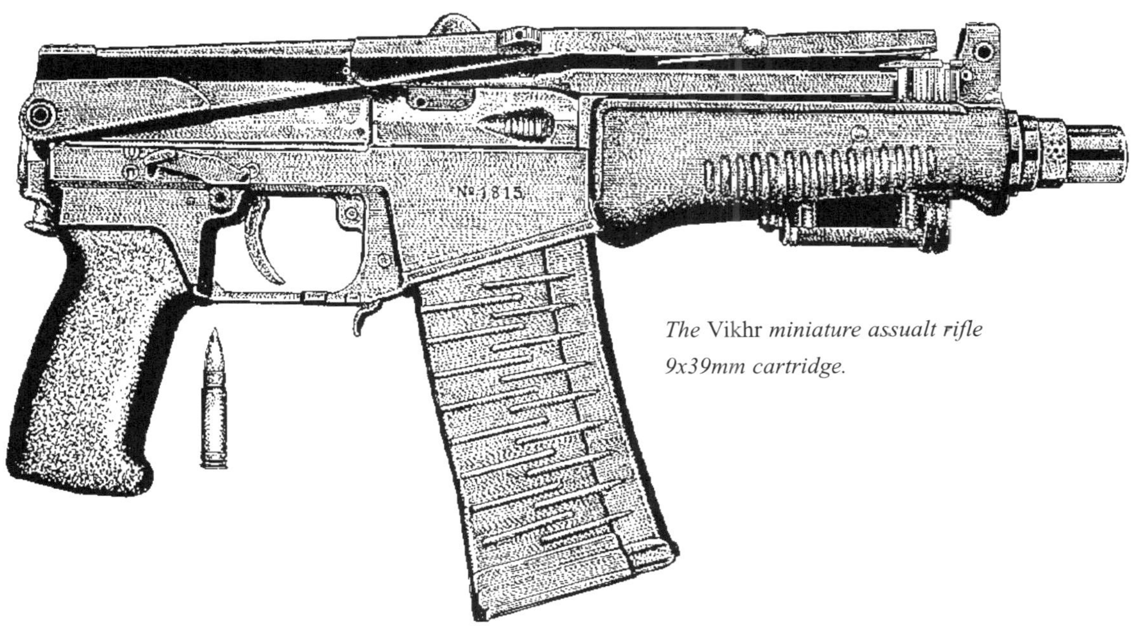

The Vikhr *miniature assualt rifle 9x39mm cartridge.*

telescopic sight, laser aiming devices, and night-vision optics. It should be noted that the PSO-1 telescopes all seemingly carry the same designation, but I have observed some that have different reticles, and the PSO-1 used on the 12.7x108mm V-94/OSV-96 (see p. 43) is 13x, rather than the standard 4x. The A-91 probably accommodates the same range of optics as the VSK-94 rifle discussed above. The A-91 also comes equipped with a suppressor that is, according to Russian sources, specifically designed for the A-91. The suppressor is a "screw-on" design and is claimed to achieve a noise reduction level of 20 decibels. The life of the suppressor is not stated. The A-91 incorporates a standard compensator that must be removed by loosening a retaining nut prior to installing the suppressor. The A-91 will also accept GP-25, GP-30, and GP-95 underbarrel grenade launchers.

A-91 SPECIFICATIONS

Caliber	9x39mm (also available in 7.62x39mm, 5.45x39mm, and 5.56x45mm)
Operation	Gas, Select Fire
Weight	3.85 lbs. (1.75 kg.)
Length, Stock Extended	23.7 in. (604mm)
Stock Folded	15.2 in. (384mm)
Barrel Length	App. 12 in. (304mm)
Effective Range	200 m.
Cyclic Rate	700–900 RPM
Magazine Capacity	20 Rounds
Sights	Front, Protected Blade; Rear, Notch. Optical

MA *VIKHR* (WHIRLWIND) COMPACT ASSAULT RIFLE

The MA is a product of TsNIITochmash and is a direct competitor with the previously described Tula A-91. Both are gas-operated, select-fire, compact assault rifle/carbines in 9x39mm caliber. The MA has two derivatives, also previously discussed, the AS silent assault rifle and the VSS silent sniper rifle.

The MA and all its derivatives are in Russian military and MVD service, a fact that has been confirmed by observation. It is possible that the MA competed with the A-91 and won, given that the MA has been observed in the hands of Interior Ministry troops and the A-91 has not, although Tula KBP claims that the A-91 is also used by MVD forces. Like the A-91, the MA is chambered for the 9x39mm SP5/SP6 cartridge. There is no evidence that the MA is available in any other caliber. TsNIITochmash brought the MA and other firearms to the United States for a demonstration, which I attended, and its representatives indicated that the MA had not been manufactured in any other calibers. Oddly, however, in Russian small arms publications, the AS, VSS, and A-91 are discussed and illustrated, but the MA is not. The reason for this anomaly is not known.

The MA's gas system is virtually identical to that of the basic Kalashnikov design, but, like that of the A-91, is shortened for use in a compact, short-barreled rifle. It is probable that the gas system of the AKS-74U served as the model for both the MA and the A-91, because both weapons are very similar to the compact Kalashnikov in design and appearance.

I have physically examined only the MA, but from detailed photographs, it appears that the MA is a slight improvement over the A-91 in terms of simplicity and execution. Quality of finish and fit on the MA seems to be slightly better than that of the A-91, but this does not necessarily affect the function of the weapon. All told, the differences between the two weapons are so slight that it is difficult to determine the selection process that the Russian government used to settle on one as opposed to the other, if this is indeed the case.

As can best be determined there is no suppressor for the MA; those desiring a weapon with a suppressor are apparently expected to use the similar AS silent assault rifle. Unlike the basic Kalashnikov and the A-91, the MA is striker fired. Feed is from 10- or 20-round translu-

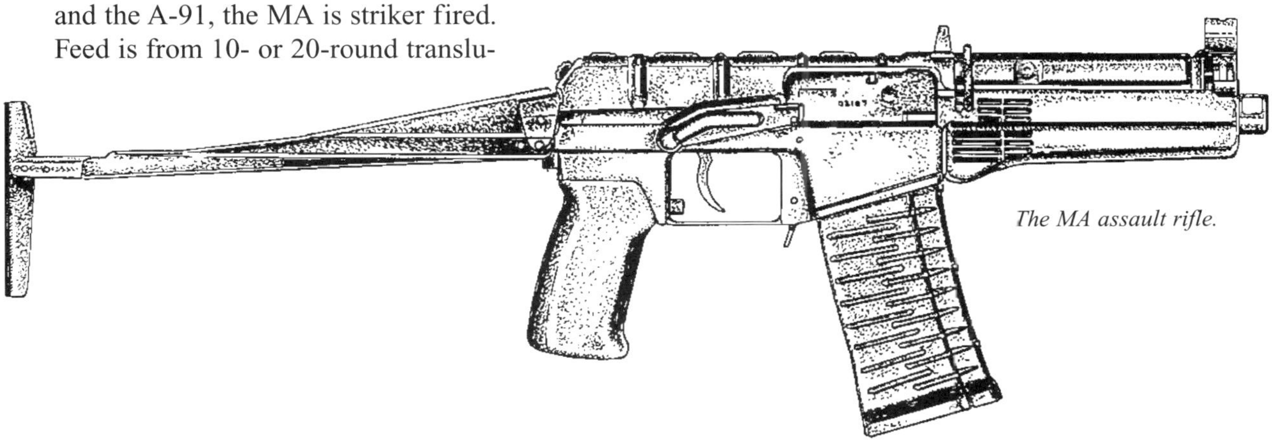

The MA assault rifle.

cent phenolic magazines. The MA receiver cover is not directly removable; it is hinged at the forward end like that of the AKS-74U and lifts upward to allow access to the operating mechanism. The cover is released by a rear latch similar to that of the basic Kalashnikov. The selector switch is ambidextrous.

As mentioned previously, the MA has been shown in the United States, and it is possible that a few have been sold to U.S. military forces for test and evaluation and for training.

MA *VIKHR* SPECIFICATIONS

Caliber	9x39mm
Operation	Gas, Select Fire
Weight	4.4 lbs. (2 kg.)
Length, Stock Extended	24.4 in. (640mm)
Stock Folded	14.9 in. (380mm)
Barrel Length	App. 10 in. (254mm)
Effective Range	200 m.
Cyclic Rate	App. 750 RPM
Magazine Capacity	20 Rounds
Sights	Front, Protected Blade; Rear, Protected Notch, Adjustable

APS UNDERWATER ASSAULT RIFLE

The APS is one of the best examples of the secrecy that surrounded the Soviet Union, and to some extent still surrounds Russia, because the rifle has been in active military service for more than 20 years and until its manufacturer, TsNIITochmash, began advertising it in 1993, was completely unknown in the West, even at classified levels. Its companion SPP-1 underwater pistol (see Chapter 2) was likewise unknown until it appeared on the market at the same time as the APS. Both weapons are unique. The German firm Heckler & Koch produces a specialized underwater pistol, the P11, but it is similar to the SPP-1 only in basic concept. No known Western manufacturer produces an underwater rifle of any type.

The APS was designed by a team headed by Vladimir Simonov, the nephew of Sergei Simonov, designer of the ubiquitous SKS carbine. The younger Simonov's team also designed the SPP-1 pistol discussed below. The design was in response to a requirement in the 1970s from the Soviet military for an underwater weapon that could be used against enemy divers. The APS is a smoothbore weapon. The bullets are actually long, thin rods with a length-to-diameter ratio of approximately 21 to 1. They are stabilized by water flow along the sides of the long, thin rod/bullet as it passes through the water rather than by spinning. The details of ammunition for the APS are discussed in the ammunition chapter of this book.

The designers began with the Kalashnikov rifle as the basis for their work, but quickly found that although the AK would work under water, the bullets had virtually no effective range and suffered from gross inaccuracy. The result of much experimentation and design work was the APS, which was fielded in the mid-1970s. The rifle is gas operated and select fire and feeds from a detachable box magazine. The odd shape of the magazine is an accommodation for the very long

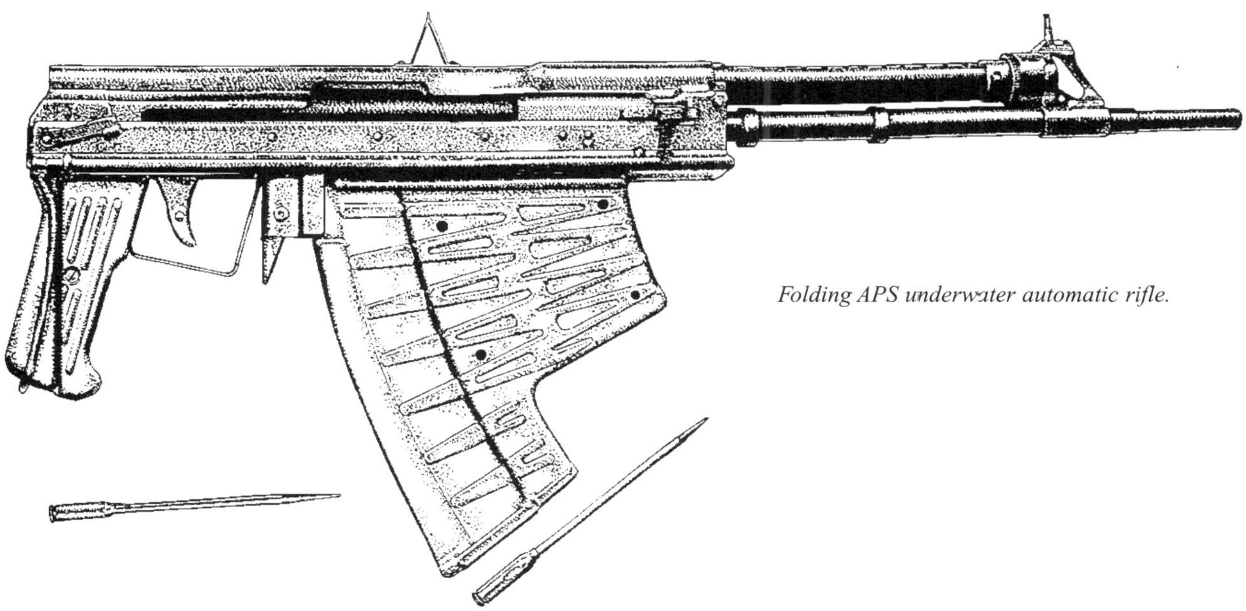

Folding APS underwater automatic rifle.

MPS cartridges necessitated by the length of the steel bullets. The APS fires from the open bolt to allow the barrel to fill with water, which is necessary to its reliable operation with the rod-shaped bullets it fires.

The range of the APS is dependent on the depth at which it is fired: the deeper, the shorter the effective range. Still, the range and lethality of the APS is greater than any spear gun, regardless of depth. The bullet of the APS does not effectively stabilize in air, so its lethal range out of water is less than 100 meters. Accuracy even at this range is questionable, given that the long bullet is unstable from the moment it leaves the barrel of the rifle.

The only operational employment of the APS that the Russian government has made public is its use during meetings between U.S. President George Bush and Russian Premier Mikhail Gorbachev at Malta in November 1989. The ships on which the two heads of state met were protected by a 16-man team of divers armed with both the APS underwater assault rifle and SPP-1 underwater pistol. This team worked in shifts and provided 24-hour watch in the waters surrounding the vessels. Their orders were to shoot to kill anyone who approached within 200 meters of the ships.

It is not known how many, if any, APS rifles have been sold on the international arms market to Western users. The rifle has, however, been in active service with Soviet and Russian special-operations units for many years, and it does provide a unique capability. TsNIITochmash has been aggressively marketing the weapon, so it may be assumed that the rifle is also in service outside Russia. Special-operations users of such weapons, however, are reluctant to openly announce what weapons are hidden in the bowels of their arms rooms, so we should not look for public announcements from Western special forces to the effect that they have adopted the APS—even if they have.

APS SPECIFICATIONS

Caliber	5.66x39mm
Operation	Gas, Select Fire
Weight	5.94 lbs. (2.7 kg.)
Length, Stock Extended	32.4 in. (823mm)
Stock Folded	24.2 in. (614mm)
Barrel Length	App. 20 in. (508mm)
Effective Range	10–30 m. Underwater, Depending on Depth; 100 m. in Air
Cyclic Rate	350 RPM
Magazine Capacity	20 Rounds
Sights	Front, Blade; Rear, Notch

12.7x108mm V-94 AND OSV-96 HEAVY SNIPER RIFLES

The V-94/OSV-96 sniper rifle is similar in concept to the many weapons that have been fielded by Western military forces in the past 10 years. Although I have categorized the two rifles as separate in the heading, the only significant differences between the V-94 and OSV-96 are the stock and the carrying handle. For that reason, I will treat both the V-94 and OSV-96 as a single rifle for the purposes of this book.

Stated weight, effective range, and dimensions of both rifles are identical. The V-94/OSV-96 is manufactured by the Tula KBP and has a claimed effective range of 2,000 meters. The rifle is gas operated and has a large muzzle brake for recoil reduction, along with a padded buttstock. The barrel folds to the right rear for carrying in the field, a virtual necessity given the 1,100mm length of the barrel. The barrel is supported for firing by a bipod approximately one-third of the way forward of the receiver. The bipod appears to be essentially similar to that used on the RPK- and PK-series machine guns. The carrying handle is adjustable for carry of the rifle "at the balance" whether the barrel is folded or extended.

Like its Western counterparts, the V-94/OSV-96 is intended as an antimateriel weapon. Russian literature indicates that the V-94/OSV-96 is intended for use against "protected personnel, lightly armoured vehicles, radars, missile launcher and artillery systems, landed aircraft . . . coast protection against small surface

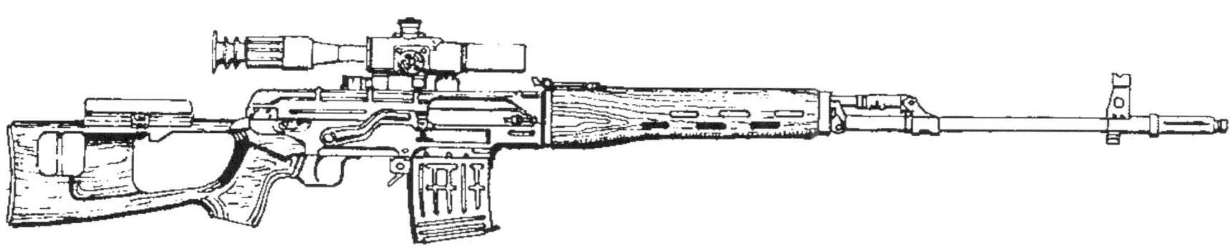

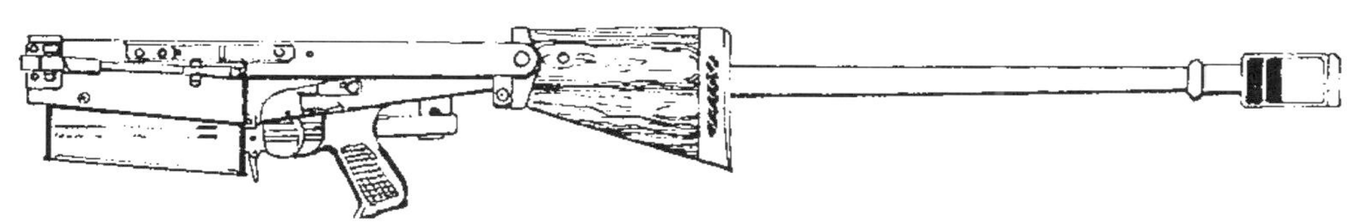

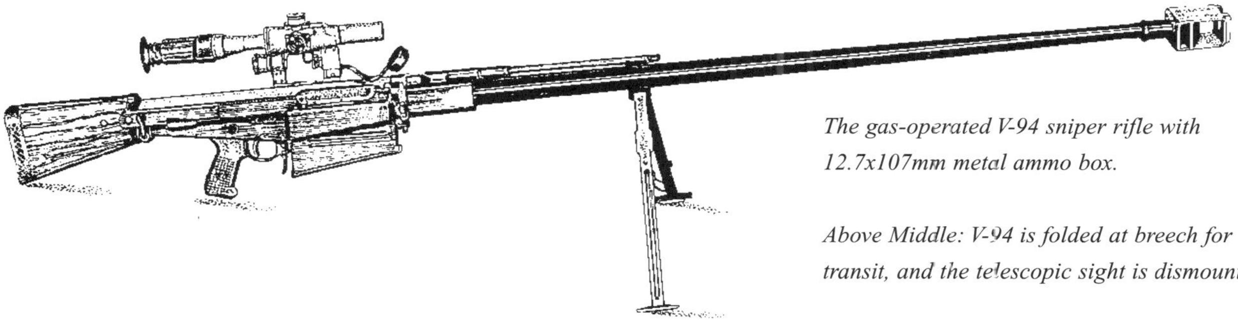

The gas-operated V-94 sniper rifle with 12.7x107mm metal ammo box.

Above Middle: V-94 is folded at breech for transit, and the telescopic sight is dismounted.

ships," and a variety of other material targets. The Russians make the point that weapons like the V-94/OSV-96 are very cost-effective and can substitute for light antiarmor weapons in certain situations.

The V-94/OSV-96 is equipped with a modified version of the PSO-1 4x sniper telescopic sight used on the SVD Dragunov rifle. Although externally identical in appearaence to the PSO-1, the modified version used with the V-94/OSV-96 is actually 13x, more in keeping with the effective range of the V-94/OSV-96 system.

The V-94/OSV-96 can fire any 12.7x108mm cartridge, but there is a special sniper cartridge designed expressly for use in the rifle. Other than its existence, there has been little in the Russian press regarding the new round. Presumably the 12.7x108mm sniper cartridge is manufactured to a higher level of precision than standard ammunition in much the same way that match ammunition is produced. The 12.7x108mm duplex round would also be a useful cartridge for the V-94/OSV-96.

The V-94 caused a minor panic in the United Kingdom in 1996 when intelligence reports were circulated that the Irish Republican Army (IRA) had acquired two of the Russian "super guns." The notion that the IRA had a rifle capable of engaging point targets out to 2,000 meters had the British press in an uproar and the police scurrying about trying to increase the security of government personnel and the royal family. Fortunately, the rifles were never used, if the IRA indeed ever had them.

V-94/OSV-96 SPECIFICATIONS

Caliber	12.7x108mm
Operation	Gas, Rotating Bolt
Weight	25.7 lbs. (11.7 kg.)
Length, Stock Extended	66.9 in. (1,700mm)
Stock Folded	43.3 in. (1,100mm)
Barrel Length	40.1 in. (1,020mm)
Effective Range	1,829 yds. (2,000 m.) (600 m. Using Night Vision Optic)
Cyclic Rate	350 RPM
Magazine Capacity	5 Rounds
Sights	13x PSO-1 Variant, Designation Unknown

6mm "UNIFIED" ASSAULT RIFLE AND MACHINE GUN

These two weapons are included together because they have never appeared separately in Russian literature or at arms shows. The two weapons were developed by TsNIITochmash and began appearing at international arms shows in 1994, but have not been shown in 1997, although literature is still being distributed in an apparent effort to find a Western manufacturer willing to enter into a joint venture to manufacture the weapons (an unlikely prospect). Given the performance of these two weapons, which shows levels of claimed effectiveness similar to those of the AN-94, it can be inferred that they were perhaps losers in the Abakan competition, which led to the adoption of the AN-94. This has yet to be confirmed by Russian sources; representatives of TsNIITochmash, the developer of both these weapons, were unwilling to discuss them when questioned by the author, suggesting that they are no longer interested in pursuing the sale of the weapons. Nonetheless, we will include them here because they are interesting developments in the new world of Russian small arms.

If these weapons were, indeed, rivals in the Abakan competition, the question arises as to why they were not adopted. Both weapons are essentially modified versions of the venerable Kalashnikov design and thus would have been much easier to introduce into the Russian military because training troops on the new weapons would have been easy. In fact, the weapons achieve their increased effectiveness through more powerful cartridges, which would have defeated one reason for adopting the AN-94—reduced recoil—unless these weapons incorporate a very efficient buffer system and muzzle brake. Moreover, a new small-arms cartridge was something that the Russian military definitely did not require. It takes many years to make the transition from one weapon to another, and if these had been adopted, the Russians would have had yet another small-arms cartridge caliber to further complicate an already complex ammunition logistics system, whereas the AN-94 is in standard 5.45x39mm caliber.

According to Russian sales literature, the Unified weapons offer effectiveness increases over 7.62x39mm weapons as follows:

Offhand (Standing)	5.1 times
Prone unsupported	1.5 times
Prone supported	2.2 times

These are no better than those of the AN-94, which uses the standard Russian military cartridge. The term *unified* probably refers to the unified caliber of the rifle and machine gun, but this is speculative.

Little information has been made public about either the rifle or the machine gun, although both are based on current designs. The rifle is clearly derived from the Kalashnikov and appears to be based on an experimental Kalashnikov rifle designated the AL-7 that was designed in the late 1960s in an effort to reduce recoil and muzzle climb on full-auto fire. The design was successful and was even certified as acceptable for military service, but never entered production. A comparison

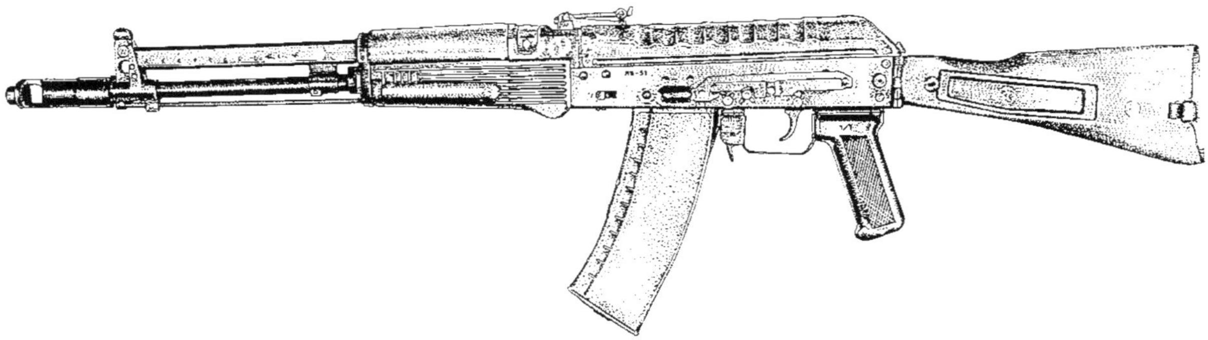

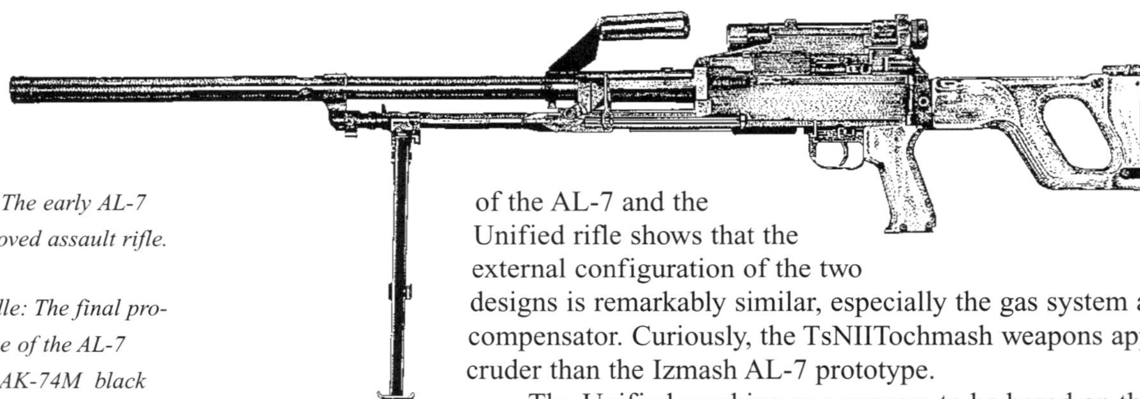

Top: The early AL-7 improved assault rifle.

Middle: The final prototype of the AL-7 with AK-74M black polyamide furniture.

Bottom: The 6mm Unified machine gun.

of the AL-7 and the Unified rifle shows that the external configuration of the two designs is remarkably similar, especially the gas system and compensator. Curiously, the TsNIITochmash weapons appear cruder than the Izmash AL-7 prototype.

The Unified machine gun appears to be based on the PK/PKM machine gun with a heavier, longer barrel. There is only a slight weight advantage over the basic PK or PKM machine guns; in fact the 14.3-pound (6.5-kilogram) Unified prototype is only 5 pounds lighter than the 20-pound (9-kilogram) PK. The advertising data for the new machine gun claim a 30 percent reduction in ammunition weight, decreased recoil momentum, and smaller overall cartridge dimensions, while achieving an increase in firing effectiveness of 2 to 2.5 times. The data are typically vague; there is no standard against which to compare them, nor is it stated how they were determined. Whatever the advantages of this prototype, they apparently were insufficient for it to be adopted by the Russian military, and TsNIITochmash began offering both it and the Unified rifle for sale or joint production with foreign partners when the decision was taken to adopt the AN-94 in 1994. TsNIITochmash literature I obtained in late 1997 still showed both weapons, now chambered in either 5.45x39mm or 5.56x45mm. There was no reference to the 6x49mm cartridge. As of early 1998, there have been no takers for the weapons in any caliber, and it appears that they will probably never enter production, as they do not offer sufficient improvements over existing weapons to warrant the expense of developing the tooling to manufacture them.

UNIFIED RIFLE AND MACHINE GUN SPECIFICATIONS

	Unified Rifle	**Unified Machine Gun**
Caliber	6x49mm	6x49mm
Operation	Gas, Select Fire	Gas, Automatic
Weight	7.04 lbs. (3.2 kg.)	14.3 lbs. (6.5 kg.)
Length	347 in. (940mm)	45.2 in. (1,150mm)
Barrel Length	App. 20 in. (508mm)	App. 28 in. (711mm)
Effective Range	600 m.	1,500 m.
Cyclic Rate	App. 700 RPM	App. 700 RPM
Magazine Capacity	30 Rounds	NA, Belt Fed
Sights	Front, Blade; Rear, Notch	Front, Blade; Rear, Notch

CHAPTER TWO

PISTOLS AND REVOLVERS

There have been almost as many new pistols and revolvers introduced or made public by the Russians in the past five years as during the country's entire history. There is a great deal of data available for some, much less for others. Some are obvious efforts by the Russian arms factory involved to regain market share through partnership with Western firms, while others are weapons that have already entered Russian military or police service and are now being offered to other customers. As with rifles, the pistols and revolvers discussed here are not intended to be an all-inclusive compendium of Russian handguns. Our focus is on special-purpose and innovative weapons that have not previously been reported on in any detail in the West. Thus the well known Makarov-based PB and Stechkin-based APB are not present, although both are special-purpose suppressed pistols, while the thoroughly conventional 5.45x18mm PSM *(Pistolet samozaryadniy malogabaritniy*, miniature semiautomatic pistol) is, along with one or two extremely unusual pistols that have been in service for more than 20 years.

Handguns of all types are now "unwelcome guests" in many countries, and those of Russian manufacture are no exception, especially those that are specifically designed to penetrate body armor, that are completely silent (as opposed to suppressed), or that are so small as to be readily concealable. Indeed, the United States cut off the importation of all Makarov-designed pistols, primarily because they are inexpensive, easily concealable, and thus desirable for personal defense by citizens of modest means. It is a well-known fact that criminals do not generally buy their weapons, but rather steal them. Given its continuing popularity and overall excellence of design, I will begin this chapter with developments based on the Pistolet Makarovka (PM). I will not address the original PM in any depth because it has been well documented elsewhere and does not meet the criteria for inclusion. I will, however, examine several interesting derivations of it, some of which were destined for the U.S. market before being banned for political purposes.

PISTOLET MAKAROVKA AND ITS DERIVATIVES

The PM was developed as a replacement for the Tokarev TT-33 7.62x25mm pistol and was adopted by the Soviet Union in 1951. The idea behind the PM was to develop a pistol that was easier to control on follow-up shots than the TT-33. Candidates for the competition conducted in the late 1940s had to be smaller and lighter than the TT-33 and be chambered for 7.62mm or 9mm cartridges. Curiously, Russian sources do not specify which 7.62 or 9mm cartridges were designated, but presumably the 7.62 was the standard 7.62x25mm while the 9 was the then-new 9x18mm. Makarov had developed and submitted prototypes in both calibers, but his 9x18mm candidate won the competition.

The PM is based on the Walther PP design of the 1920s, but improves upon it in several ways. First, the PM has an external slide stop that enables the user to insert a fresh magazine and then release the slide with one hand by simply pressing down on the release. The Walther requires that the user pull the slide to the rear and release it, which is far more difficult, especially in combat situations. Second, the PM safety works in the reverse of that of the Walther pistols. The Walther safety is "on" when it is in the "up" position, requiring the user to make a generally unnatural move to release it by reaching up with his thumb and then bringing the switch down to the "off" position. The PM safety is "on" when down and moving it to the "off" position is as simple as flicking one's thumb forward. The one change that Makarov made that detracts from the design of his pistol to most users is the relocation of the magazine release from behind the trigger guard to the base of the butt, a real detriment to quick reloading.

Another shortcoming of the PM is its ammunition. Although the 9x18mm full-metal jacket (FMJ) "ball" round has better external and terminal ballistics than the 9x17mm (.380ACP) FMJ round, it is not nearly as effective as the 9x19mm (9mm Lüger or Parabellum) FMJ cartridge. The 9x18mm is the most powerful cartridge available anywhere in a direct blowback-operated firearm, but it is not a very effective combat load. The Russian military probably reached this conclusion some time ago, which has led its arms designers down some very interesting paths.

The first of these is a new round for the PM, the 57-N-181-M, which retains the 9x18mm cartridge case but raises the muzzle velocity from 315 meters per second (800 fps) to 410 meters per second (1,041 fps). A simple calculation reveals

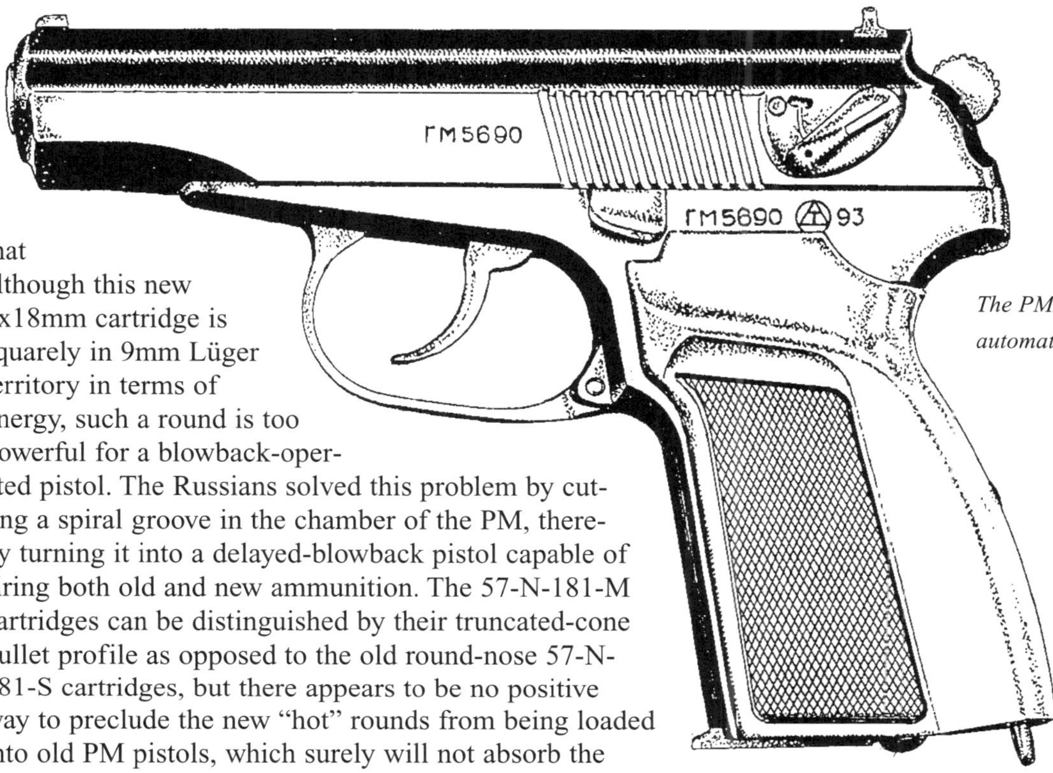

The PMM-8 9x18mm automatic pistol.

that although this new 9x18mm cartridge is squarely in 9mm Lüger territory in terms of energy, such a round is too powerful for a blowback-operated pistol. The Russians solved this problem by cutting a spiral groove in the chamber of the PM, thereby turning it into a delayed-blowback pistol capable of firing both old and new ammunition. The 57-N-181-M cartridges can be distinguished by their truncated-cone bullet profile as opposed to the old round-nose 57-N-181-S cartridges, but there appears to be no positive way to preclude the new "hot" rounds from being loaded into old PM pistols, which surely will not absorb the higher chamber pressures and concomitant increase in slide velocity for very many shots. Russian sources state that the dimensions of the new cartridge are identical to those of the old, so this must surely constitute something of a problem in the Russian military and MVD forces, which are equipped with weapons that chamber both old and new 9x18mm cartridges. As this was written, several new Russian pistols and submachine guns have appeared in 9x19mm (9mm Parabellum), indicating that the Russians may not be completely satisfied with the new high-impulse 9x18mm cartridge.

The new delayed blowback pistol was designated the PMM, or Pistolet Makarovka Modernizirirovannyi. It is more than simply a PM with a spiral groove in the chamber, however. In addition to firing the new, more powerful 9x18mm cartridge, the PMM has been modified to accept a 12-round, staggered box magazine. This version is officially designated the PMM-12. An eight-round version of the PMM was also developed. This version is designated simply PMM. Both pistols are about 5 percent larger than the basic PM. Most Russian literature describes the PMM-12 rather than the eight-round version, and this is probably the more common of the two pistols. Most Russian literature also refers to the 12-round pistol as PMM, so that is the terminology we will use here to describe the high-capacity pistol.

The Russians directly compare the PMM and its improved 9x18mm round with Western 9x19mm designs, such as compact versions of Glock, SIG, Smith & Wesson, and Heckler & Koch pistols. The Russians point out that the PMM delivers equal or better terminal ballistics than many 9x19mm loads, but in a smaller

Above: The PMM-12 9x18mm automatic pistol with 12-round magazine.

Below: The OTs-35 is a compensated Makarov.

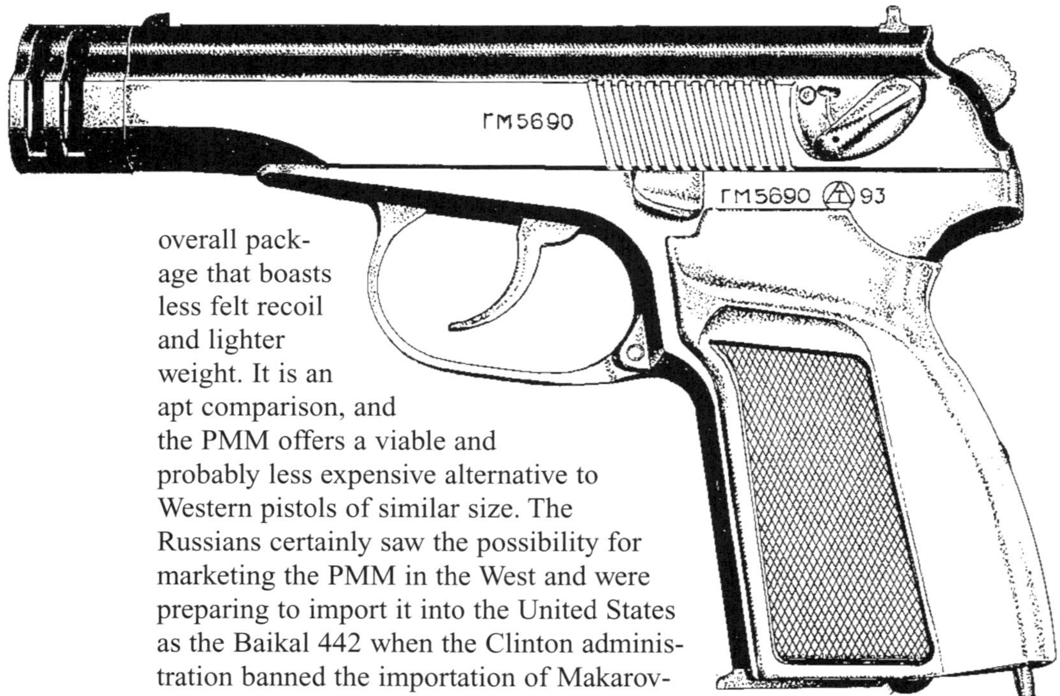

overall package that boasts less felt recoil and lighter weight. It is an apt comparison, and the PMM offers a viable and probably less expensive alternative to Western pistols of similar size. The Russians certainly saw the possibility for marketing the PMM in the West and were preparing to import it into the United States as the Baikal 442 when the Clinton administration banned the importation of Makarov-type pistols.

The Baikal 442 version of the PMM that would have been imported into the United States would not, however, have been chambered for the new 9x18mm cartridge, but for the 9mm Lüger. The forces generated by the 57-N-181M 9x18mm cartridge are similar to those of the 9mm Lüger, and redesigning the PMM to accept the slightly longer (4mm overall) 9mm Lüger cartridge was not a great engineering challenge. The illustration of the Baikal 442 (on p. 57) clearly shows the intent to market the pistol in the United States. This pistol has fully adjustable sights and a relatively high level of finish in contrast to earlier pistols. The Baikal 442 is also slightly larger than the PMM manufactured for domestic Russian use to ensure that it is in consonance with U.S. import standards. The Baikal 442 is still in production at Izmash, but it will unfortunately not be imported into the United States.

The PMM will probably eventually replace the older PM in Russian military and police service, because it represents a significant improvement over the PM. Moreover, the ability to chamber and fire the new high-impulse ammunition in older pistols is an obvious problem that can only be completely rectified by changing over to the PMM. Indeed, the PMM has been in series production by Izmash since January 1994 and could eventually replace the PM in the Russian military.

A final variant of the PMM Makarov is the OTs-35 manufactured by Tula KBP. The OTs-35 is essentially a PMM Makarov with a recoil compensator added to the muzzle. Tula sources state that the reason for adding the compensator was to increase accuracy and hit probability by reducing muzzle rise, especially when firing the high-impulse PMM cartridge. Weight increase is negligible over that of the basic PMM; the compensator adds an extra 18mm (approximately 3/4 inch) to the overall length of the pistol.

PMM (PM) SPECIFICATIONS

Caliber	9x18mm Mod (9x18mm)
Operation	Delayed Blowback (Blowback), Semiautomatic
Weight	26 oz. (.76 kg.) (24 oz./.68 kg.)
Length	6.6 in. (167mm) (6.4 in./161mm) 28.6 in. (728mm)
Barrel Length	3.6 in. (93.5mm)
Effective Range	25 m.
Magazine Capacity	12 Rounds* (8 Rounds)
Sights	Front, Blade; Rear, Notch, Adjustable for Windage

*An eight-round version of the PMM is also available.

PSM "MINIATURE SELF-LOADING PISTOL"

The Pistolet Samozaryadniy Malogabaritniy (miniature semiautomatic pistol), better known in the West as the PSM, is one of those Russian weapons that has been around for quite some time, but only recently has become known in Western firearms circles. The pistol has been test-fired by several Western writers, and while the characteristics of the little pistol are fairly well-known, why and how it was developed and its service use are not widely known, and it is on these aspects of the PSM that I will concentrate.

PSM design and development began in 1969 at the Tula KBP by a team led by Tikhon Lashnev. The PSM cartridge, designated the 7N7, is essentially a 9x18mm Makarov cartridge case "necked down" to 5.45mm. Despite the simplicity of the concept, its execution was extraordinary. The PSM was designed from the outset for concealed carry and to achieve high penetration by firing a bullet of high sectional density. The PSM cartridge is not only designed to achieve maximum terminal ballistics against personnel targets, but against soft body armor as well. Western tests of the diminutive round at typical CQB ranges reveal that the PSM is very effective against armored targets. When the PSM and its ammunition first began filtering into the West in the early 1990s, many European military and police organizations were alarmed at the effectiveness of the cartridge against soft body armor. Early tests showed that it would penetrate 30 to 45 layers of Kevlar with sufficient residual energy to penetrate and kill a human target. The PSM was therefore designated an "assassination pistol." Since then, Russian press articles on the little pistol have revealed a somewhat more mundane purpose for its design and fielding. The PSM was intended for use by staff officers at the highest levels of command who could not be burdened with a heavier pistol, but who required an effective means of self-defense. Since many military forces were equipping their soldiers with body armor at the time the PSM was being developed, it was only logical that the weapon be able to defeat soldiers wearing "flak vests."

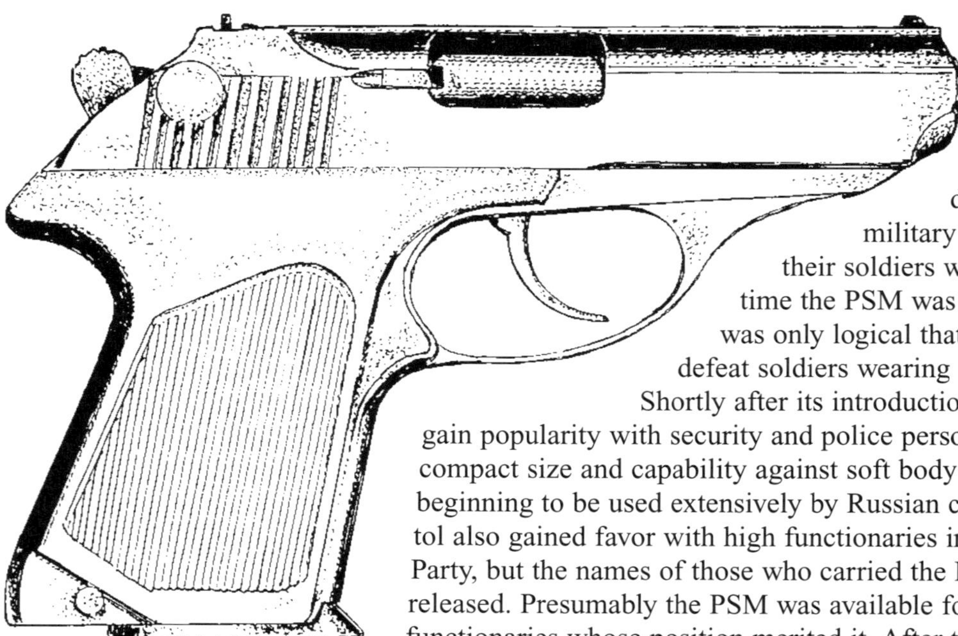

An early 5.45mm PSM automatic pistol.

Shortly after its introduction, the PSM began to gain popularity with security and police personnel because of its compact size and capability against soft body armor, which was beginning to be used extensively by Russian criminals. The little pistol also gained favor with high functionaries in the Communist Party, but the names of those who carried the PSM have not been released. Presumably the PSM was available for issue to those party functionaries whose position merited it. After the collapse of the

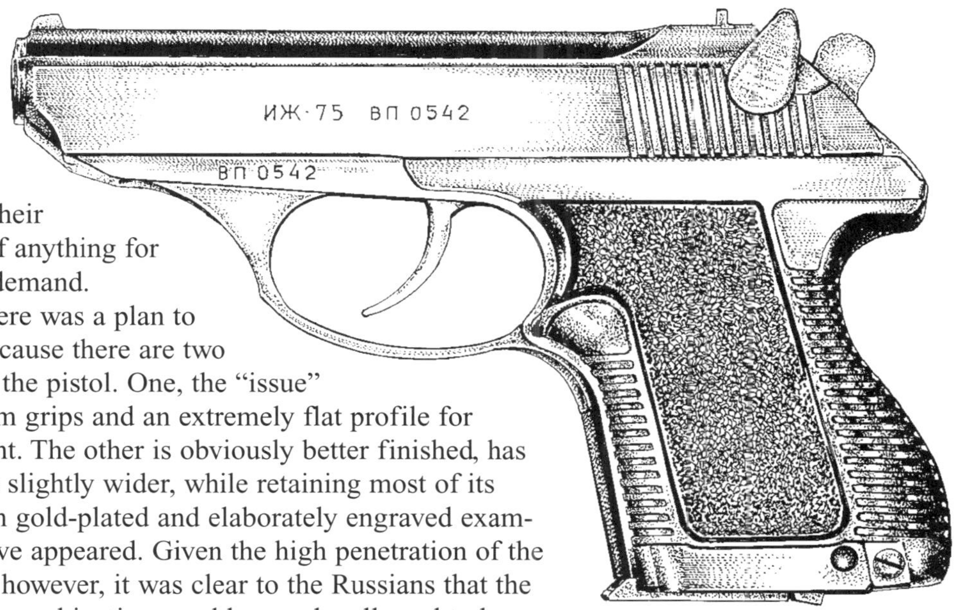

The 5.45mm PSM self-loading compact pistol.

Soviet Union, the PSM came to the attention of the Western arms market through Russian efforts to improve their economy via sale of anything for which there was a demand.

Apparently there was a plan to export the PSM, because there are two distinct versions of the pistol. One, the "issue" pistol, has aluminum grips and an extremely flat profile for optimal concealment. The other is obviously better finished, has plastic grips, and is slightly wider, while retaining most of its concealability. Even gold-plated and elaborately engraved examples of the PSM have appeared. Given the high penetration of the 5.45x18mm bullet, however, it was clear to the Russians that the cartridge and pistol combination would never be allowed to be imported into the United States, so they set out to make the PSM "politically correct" to meet U.S. import criteria. The result was the Baikal 441 pistol, which is described in the next section.

PSM SPECIFICATIONS

Caliber	5.45x18mm
Operation	Blowback
Weight	1.01 lbs. (.46 kg.)
Length	6.1 in. (155mm)
Barrel Length	3.3 in. (85mm)
Effective Range	App. 25 m.
Magazine Capacity	8 Rounds
Sights	Front, Blade; Rear, Notch

BAIKAL 441 COMPACT PISTOL

After realizing that the PSM would never meet import restrictions in Western countries, especially the United States, Izmash designers set out to make the PSM more politically acceptable to the U.S. government bureaucracy by meeting the terms of the Gun Control Act of 1968. In this they succeeded, as evidenced by the illustration of the Baikal 441, whose internal designation is IJ-75. Of course, the Clinton administration effectively ended all Russian arms imports in 1994, and despite the fact that the Baikal 441 would have been eligible for import under prior criteria, it fell victim to the antigun philosophy of the Clinton administration and thus will probably never be imported unless there is a policy change after Clinton leaves office.

Modifications to meet U.S. standards included chambering the Baikal 441 for the .25 ACP (6.35mm Browning) and adding larger wooden grips, a loaded chamber indicator, and adjustable sights. Otherwise, the Baikal 441 is identical to the PSM.

BAIKAL 441 SPECIFICATIONS

Caliber	.25ACP (6.35mm Browning)
Operation	Blowback
Weight	1.05 lbs. (.48 kg.)
Length	6.1 in. (155mm)
Barrel Length	3.3 in. (85mm)
Effective Range	25 m.
Magazine Capacity	8 Rounds
Sights	Front, Blade; Rear, Notch, Adjustable for Windage and Elevation

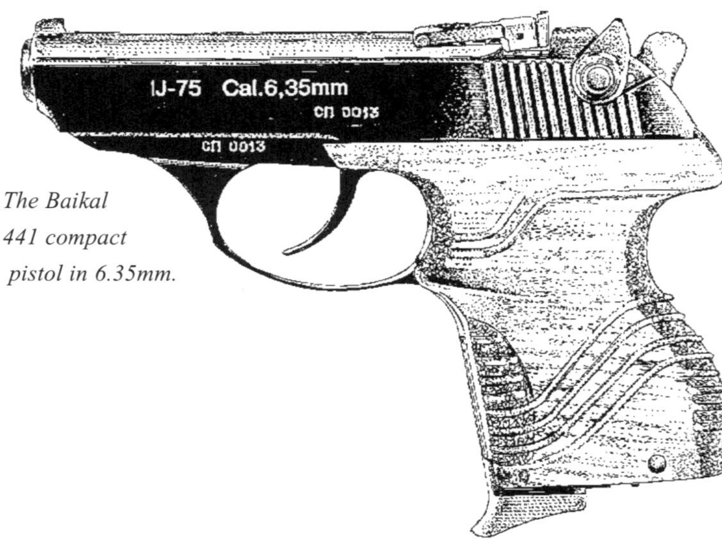

The Baikal 441 compact pistol in 6.35mm.

BAIKAL 442 PISTOL

The Baikal 442 was another attempt by Izmash to penetrate the U.S. market, and if the Clinton administration had allowed the pistol's import, it probably would have achieved excellent sales. Essentially, the Baikal 442 is a modified version of the PMM. It is still available to other markets chambered in both 9x18mm and 9x19mm cartridges. Magazine capacity is 8 rounds of 9x19mm and 10 or 12 rounds of 9x18mm. The slide weight was increased over that of the PMM, minor changes were incorporated into the frame and trigger guard, and the pistol design was altered to allow field stripping only after the magazine had been removed. This is an unusual safety feature, probably driven by liability laws in the United States, where Izmash originally intended to market the pistol.

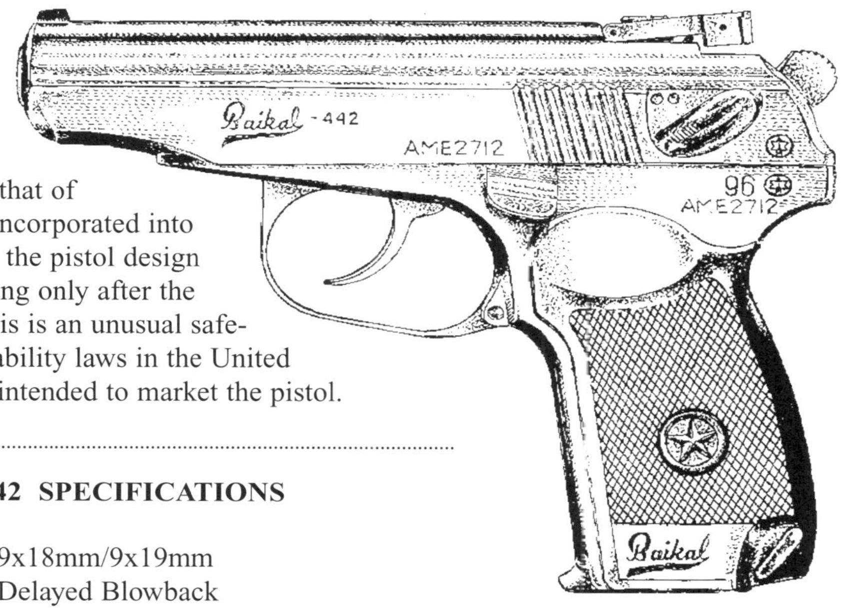

The 9mm Baikal 442 in 9x18mm or 9x19mm with retarded blowback (grooved chamber).

BAIKAL 442 SPECIFICATIONS

Caliber	9x18mm/9x19mm
Operation	Delayed Blowback
Weight	1.71 lbs. (.76 kg.)/1.6 lbs. (.73 kg.)
Length	6.6 in. (169mm)/6.33 in. (161mm)
Barrel Length	3.7 in. (93.5mm)
Effective Range	25 m.
Magazine Capacity	8, 10, or 12 Rounds/8 Rounds
Sights	Front, Blade; Rear, Notch, Adjustable for Windage and Elevation

Heretofore, we have considered pistols manufactured by Izmash and Tula KBP, two of the three major firearms producers in Russia. The third major producer is the Precision Mechanical Engineering Central Research Institute, better known as TsNIITochmash, located in Klimovsk. This firm manufactures a number of special-purpose weapons, including several pistols of unconventional design. We will begin with the most powerful, the *Gurza* or Viper.

9x21mm *GURZA* (VIPER) PISTOL

The *Gurza*, sometimes called the *Gyurza*, is unusual for the Russians in that it is a very powerful handgun; the 9x21mm Russian (as distinguished from the Western 9x21mm) Model RG-054 round is ballistically comparable with the .357 Magnum.

Further, it has completely different ballistics than does the Western 9x21mm, which was designed for use in countries that prohibited civilian ownership of military-caliber handguns and is ballistically comparable to the 9x19mm cartridge. Details of the RG-054 9x21mm Russian cartridge are in the ammunition chapter of this book.

The *Gurza* was developed to a military and police requirement for a pistol that offered a dramatic increase in terminal ballistics over the PM and other firearms chambered for the 9x18mm cartridge in all its permutations. The Russians were beginning to have serious problems with violent criminals wearing soft body armor and using high-performance European cars, such as the Mercedes-Benz, to perpetrate their crimes. Russian police thus needed a pistol capable of penetrating not only body armor but car bodies as well. According to the manufacturer, the *Gurza* is in use by the Russian police (*militsiya*) and Military Intelligence Directorate (GRU) personnel. The manufacturer also claims that the *Gurza* RG-054 bullet penetrates 2.4mm of titanium and 30 layers of Kevlar at 50 meters. I have fired this pistol against such a target, and the Russian claim is no exaggeration. Its steel-cored RG-054 cartridge represents a potent capability against body armor up through NIJ Level IIIA+ at ranges up to 50 meters.

The *Gurza* is an unconventional design in many ways. The operating system is based on that of the Beretta M92 (i.e., a falling block rather than the Browning system that appearances would indicate). The frame is part polymer and part steel rather than being all one or the other. The two components of the frame join along the line that is visible in the illustrations running along the top of the grip, above the trigger guard, and extending to the end of the frame. The pistol has no conventional safeties. There is a trigger safety that completely disengages the trigger mechanism from the sear mechanism when the hammer is completely down. Pulling the trigger when the hammer is in this position does nothing;

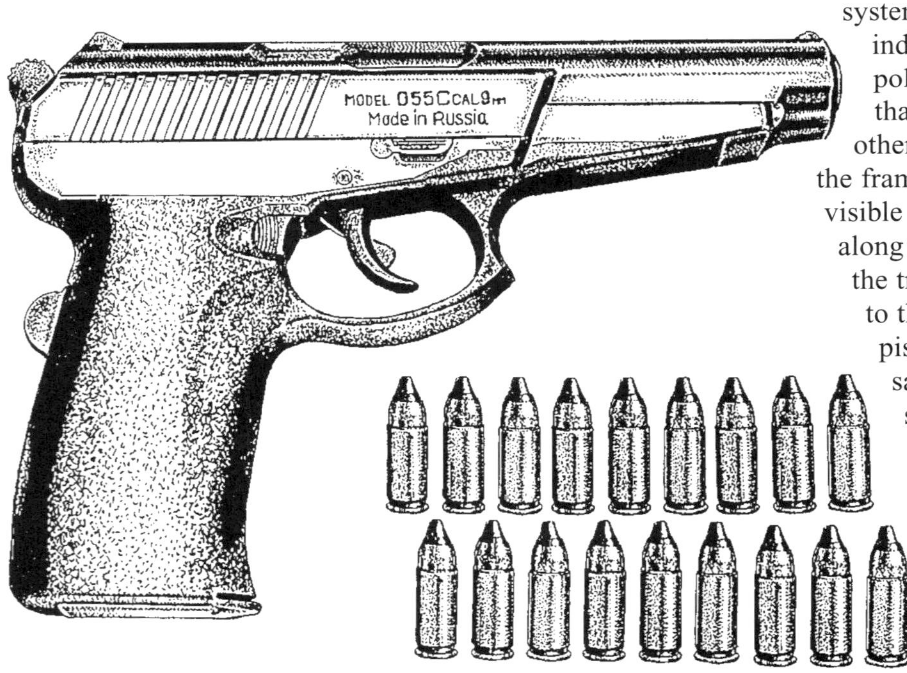

The Gurza *442 9x21mm export-model automatic pistol.*

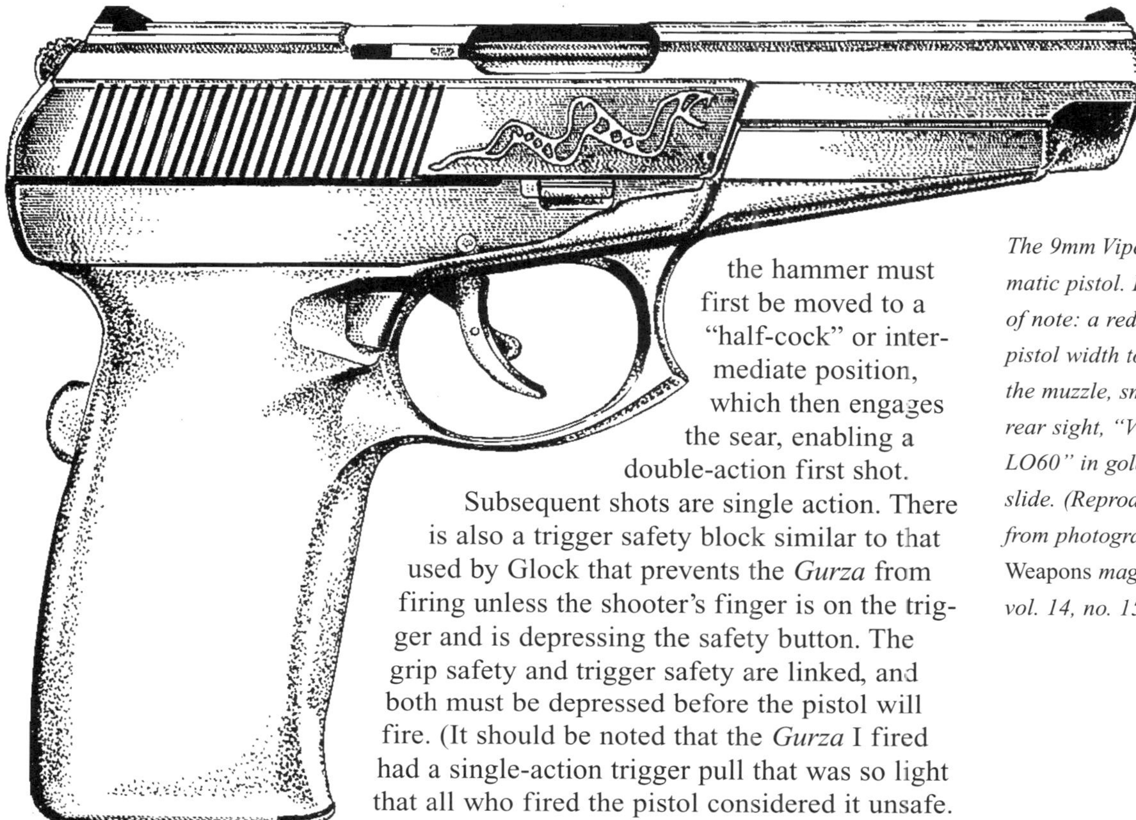

The 9mm Viper automatic pistol. Features of note: a reduction in pistol width toward the muzzle, smaller rear sight, "Viper LO60" in gold on the slide. (Reproduced from photographs in Weapons *magazine, vol. 14, no. 15.*

the hammer must first be moved to a "half-cock" or intermediate position, which then engages the sear, enabling a double-action first shot. Subsequent shots are single action. There is also a trigger safety block similar to that used by Glock that prevents the *Gurza* from firing unless the shooter's finger is on the trigger and is depressing the safety button. The grip safety and trigger safety are linked, and both must be depressed before the pistol will fire. (It should be noted that the *Gurza* I fired had a single-action trigger pull that was so light that all who fired the pistol considered it unsafe. When questioned, the Russian designers stated that the light trigger pull was a requirement by the users. Nonetheless, the single-action trigger pull was an estimated 1.5 to 2 pounds, so light that it went off seemingly the moment it was touched.) Other than these features, the *Gurza* is a fairly conventional design. The final safety feature is a grip safety, considered archaic in the West, but obviously not by the Russians. Indeed, given the design of the *Gurza*, the grip safety is necessary.

The magazine release is ambidextrous and operates with a forward push of the thumb or index finger. The *Gurza* handles extremely well, other than the trigger mechanism. Recoil was relatively easy to manage as the *Gurza* is a large pistol with wide, comfortable grips. Accuracy was good once we became accustomed to the extremely light trigger pull. As stated, the 9x21mm Russian round defeats NIJ Level IIIA+ body armor at 50 meters. All in all, the *Gurza* is unsuitable for use in the West as it is currently configured. The trigger is patently unsafe, and the lack of an external safety for use when the pistol is in single-action firing mode is a serious shortcoming, despite the trigger and grip safeties.

Even if these design faults were to be corrected, the *Gurza* would never be allowed to be imported into most countries because the only known ammunition for it is the RG-054 armor-piercing round. The pistol could very well be converted to fire other calibers, such as 10mm, .40 Smith & Wesson, or even 9x19mm, although there is no indication by either TsNIITochmash or its U.S. representatives of any inclination to do so. The basic design is good, but for it to be accepted in the West, it will have to be refined and upgraded in several areas.

GURZA 442 SPECIFICATIONS

Caliber	9x21mm Russian
Operation	Short Recoil
Weight	2.09 lbs. (.95 kg.)
Length	7.67 in. (195mm)
Barrel Length	App. 5 in. (127mm)
Effective Range	40–50 m.
Magazine Capacity	18 Rounds
Sights	Front, Blade; Rear, Notch, Adjustable for Windage

7.62x42mm PSS SILENT PISTOL

The PSS is a fairly conventional blowback semiautomatic pistol with the exception of its unique ammunition, which is silent—not suppressed. The PSS was designed by Viktor Levchenko to a special operations requirement for a compact pistol that required no bulky suppressor, and because of its silent ammunition it is very compact and readily concealable. I have fired this pistol, and the only sound that can be heard is the slide reciprocating back and forth. The PSS has no lock to prevent movement of the slide when the pistol is fired, which is unusual in a silent pistol. This is perhaps the only shortcoming of the little pistol, however. The absence of an external suppressor makes it very compact and easy to carry concealed, when desired.

The 7.62x42mm SP-4 silent cartridge is the key to the efficacy of the PSS and is covered in detail in the ammunition chapter of this book. In brief, however, the cartridge has a bullet that is flush with the mouth of the cartridge case. When the cartridge is fired, a small internal piston propels the bullet out and down the barrel. The piston obturates against an internal shoulder, trapping the propelling gases inside the case, where they are eventually allowed to slowly bleed off. The bullet itself is no more than a steel cylinder with a copper rotating band. SP-4 casings examined by the author have no headstamp to identify their origin, but, given the unique character of the 7.62x42mm cartridge, identification would pose no great dilemna.

The technology for this design was not new; it was used by the United States for special revolvers employed in the Vietnam War. According to the manufacturer, the PSS will penetrate both military helmets and soft body armor. Although we did not test the SP-4 round against a steel or Kevlar helmet, it was fired against a Level II protective vest at 25 meters, which it cleanly penetrated.

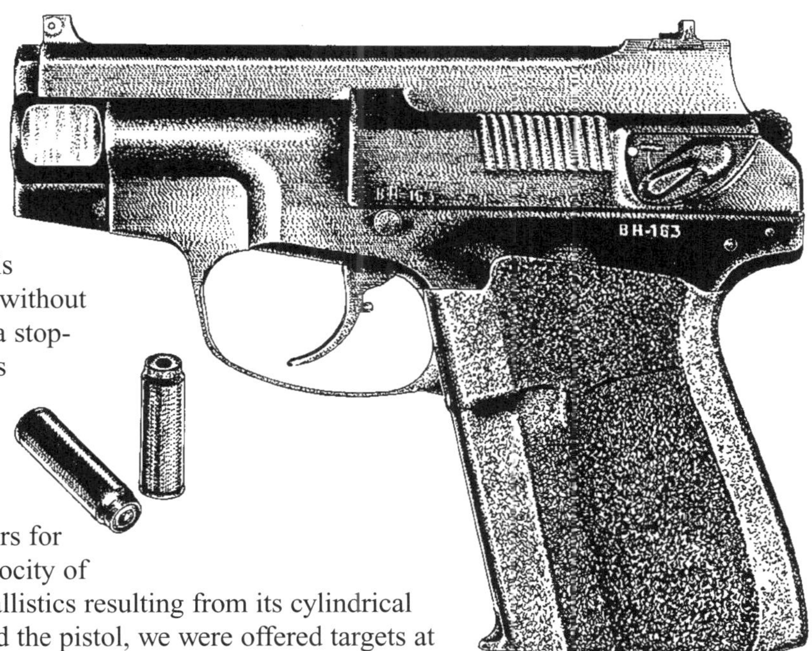

The PSS (6-P-28) self-defense pistol.

With its mild recoil, the PSS is pleasant to shoot. Functioning was without incident; the little pistol never had a stoppage in two days' shooting. There is a Makarov-type safety on the left side of the slide. The pistol is double action on the first shot and single action thereafter. The Russians claim an effective range of 50 meters for the PSS, but given the subsonic velocity of the SP-4 bullet, with its abysmal ballistics resulting from its cylindrical shape, this is doubtful. When I fired the pistol, we were offered targets at no greater distance than 25 meters, which is really more in keeping with

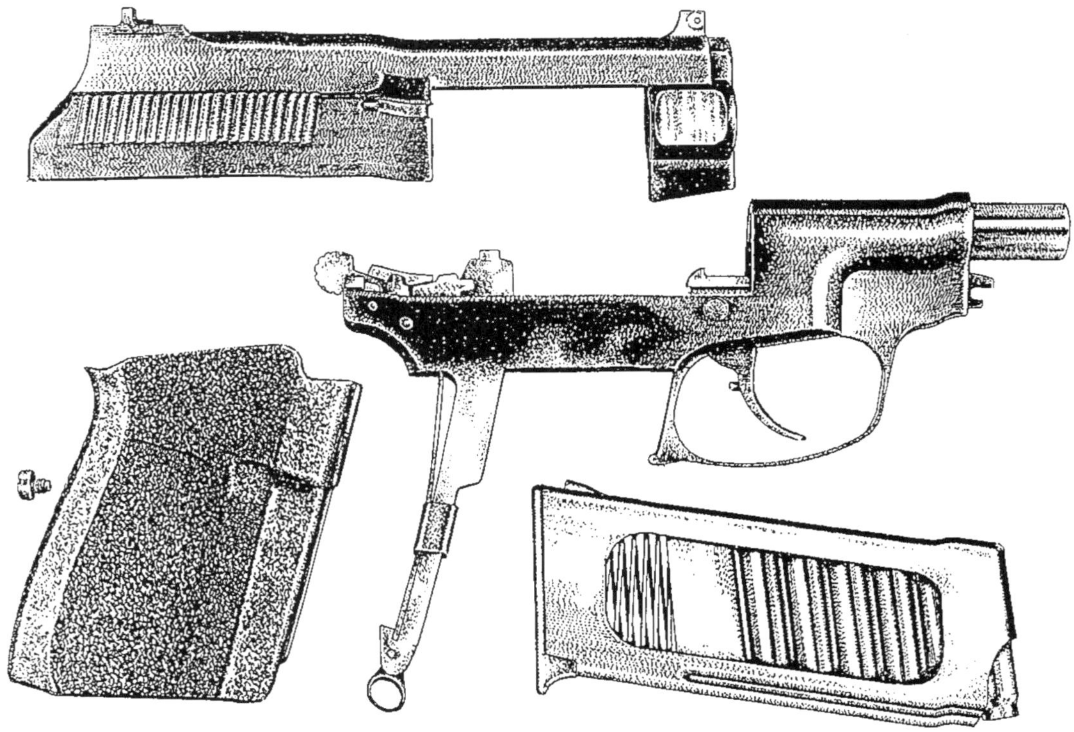

A stripped PSS.

the expected capabilities of the little PSS. For its intended purposes, however, the PSS is apparently a very effective pistol, indeed. At CQB distances of 25 meters or less and in confined spaces, its silence and lack of muzzle flash make it nearly optimal for special operations or SWAT use in special situations.

PSS SILENT PISTOL SPECIFICATIONS

Caliber	7.62x42mm
Operation	Blowback
Weight	1.54 lbs. (.7 kg.)
Length	6.5 in. (165mm)
Barrel Length	App. 3 in. (76mm)
Effective Range	25 m.
Magazine Capacity	6 Rounds
Sights	Front, Blade; Rear, Notch, Adjustable for Windage

7.62x35mm MSP SILENT PISTOL

The Malogabaritnyi Spetsial'nyy Pistolet (Small-Sized Special Pistol, MSP) was developed by TsNIITochmash in the late 1960s. It was produced by the Tula Arms Plant, and entered into service in 1972. Its Russian nickname is *Groza* (Thunderstorm), which is not to be confused with the OTs-14 rifle (see Chapter 1). The little pistol is a double-barreled "over-under" derringer-type firearm that shoots a first-generation silent cartridge. As this book was being written, there was still some confusion about the designation of the cartridge fired by the MSP.

From photographs, it can be seen that the MSP clearly fires the 7.62x35mm cartridge, which is externally similar to the 7.62x39mm M1943 round. It has been termed the 7.62x35mm to differentiate it from the earlier round. Like the SP-4 round described earlier, the 7.62x35mm round projects a bullet via a piston, which is held captive in the cartridge casing after it is fired. This round has been referred to as the SP-3, but so has another, entirely different silent cartridge that is used in the similar S4M silent pistol described in the next section. Both silent cartridges fire a bullet that is identical to the bullet fired by the AK-47 and AKM rifles. The rifling of the pistols is probably identical to that of the two rifles as well. Again, this special cartridge is covered in detail in the ammunition chapter of this book.

The MSP was developed for special operations, and even the Russians openly admit that the pistol has no useful self-defense purposes. It is clearly an assassination weapon and was used both in Afghanistan and Central America.

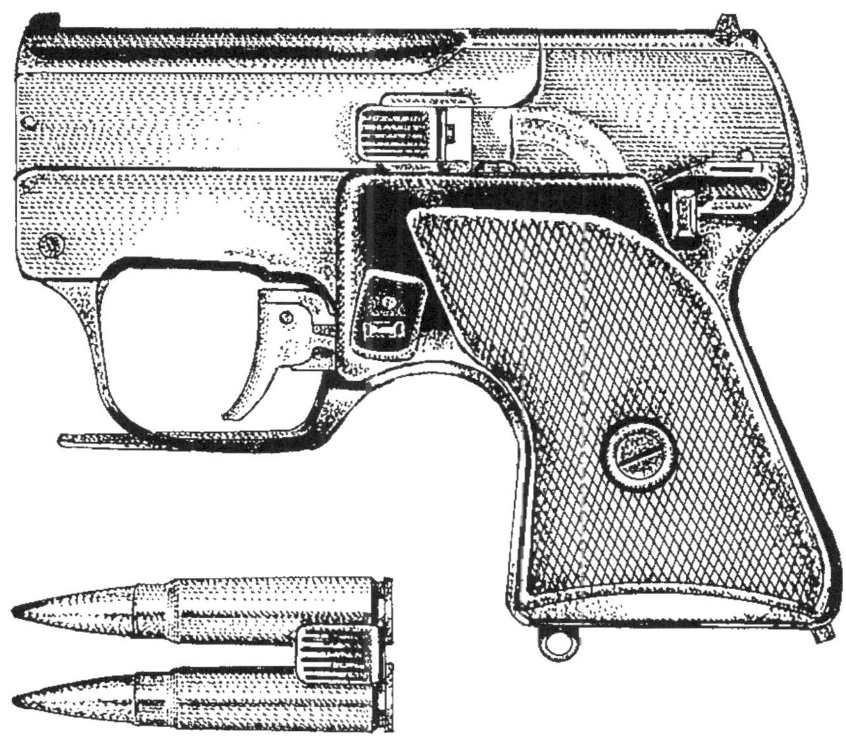

The MSP multibarrel handgun in 7.62x35mm.

For its intended purposes it was nearly ideal. The assassin simply had to approach his victim, get fairly close, and fire. There was no sound associated with the shot, because the gases were trapped inside the cartridge case and the bullet was subsonic. Moreover, there was no slide movement as there is with the current-issue PSS semiautomatic pistol. Since the bullet is the same as that fired from standard Russian rifles, its source would be a mystery to the doctors performing the autopsy on the assassin's victim, because no gunshot would have been heard.

The MSP was limited, however, in that it only fired two shots. Moreover, the walls of the 7.62x35mm cartridge case were relatively thin, which leads to the conclusion that there may well have been failures to contain the gases. The next silent cartridge, the SP-3 in 7.62x62.8mm Russian, had extremely thick case walls—even thicker than the subsequent SP-4. (The Russian cartridge is actually 63mm in length but has been designated 62.8 because the 7.62x63mm is the metric designation for the .30-06 cartridge.)

Both the 7.62x35mm and 7.62x62.8mm Russian cartridges could not be used in a semiautomatic weapon, however, because the propelling piston protruded well beyond the case mouth and up into the barrel of the pistol. (See illustrations in the ammunition chapter.)

The MSP design is simplicity itself. The barrels tip up from the rear for loading from a two-round clip. A small lever under the trigger guard is used to cock the pistol manually before firing it. Upon firing, the spent cases are extracted together, according to Russian sources. From an analysis of a Russian cutaway of the MSP, it is apparent that the spent cartridges are not thrown clear of the pistol but are lifted away from the chamber for manual extraction—leaving spent casings behind on an assassination mission would be an embarrassment not only to the assassin, but to the secret service that directed him to carry out the mission.

Although the MSP is probably obsolescent in Russian service, it is included in this volume because it is virtually unknown in the West despite the fact that it has been in service for nearly 30 years. The very similar S4M pistol described in the next section is even more mysterious, although it too has been in Russian service for many years and uses similar ammunition.

7.62x62.8mm S4M SILENT PISTOL

Like the MSP silent pistol, the S4M dates back to the 1970s, but it is a more advanced design. The ammunition used in this pistol is similar in concept to that of the MSP, but is quite different in execution. While the MSP ammunition is to all external appearances a standard 7.62x39mm round, the S4M ammunition is obviously something quite special. Again, the ammunition is covered in detail in the chapter on that subject. The S4M cartridge is also loaded in a two-round clip, and the bullet is identical to the 7.62x39mm M1943 bullet. As with the ammunition used in the MSP, the piston that projects the bullet out of the barrel protrudes from the case mouth, eliminating the potential for using the cartridge in anything but a simple, break-open pistol.

The S4M is a slightly more advanced than the MSP in several ways. It can be considered essentially a "product improved" version of the MSP. Cocking is automatic on the S4M, and the pistol has a smoother overall exterior to prevent snagging on the clothes of the assassin. Other than those improvements, little else has been made public regarding the pistol, and it remains something of an enigma.

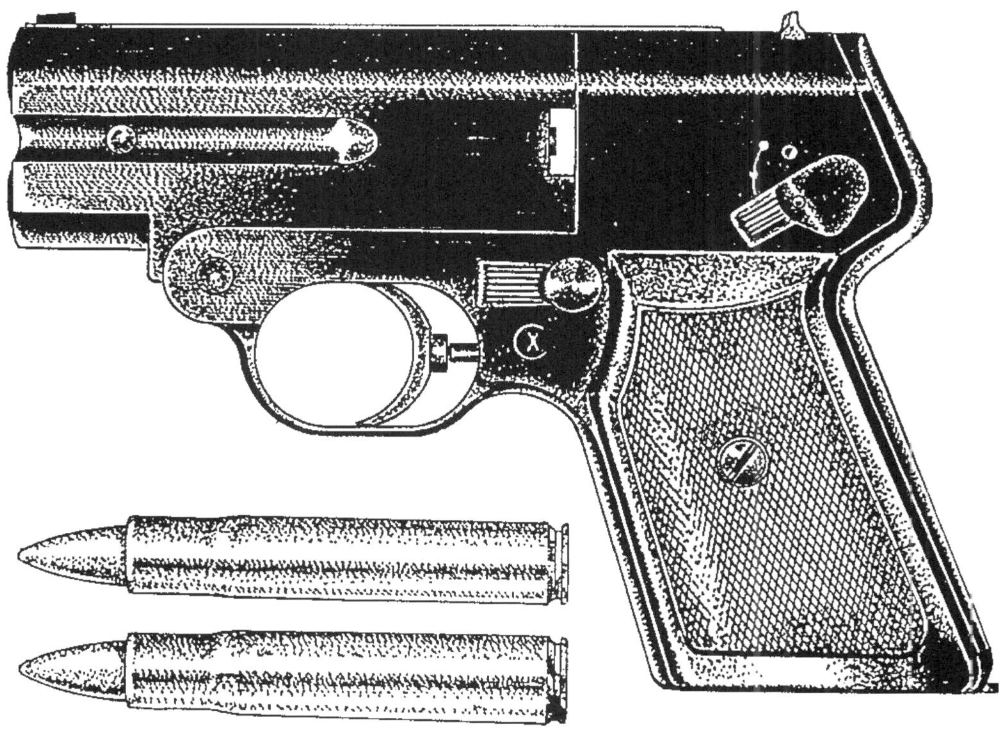

The S4M multibarrel handgun.

OTs-21 *MALYSH* (FLY) 9x18mm PISTOL

The Tula KBP-manufactured OTs-21 *Malysh* is one of several new pistols in Russia that had just been introduced into the market as this was being written. The Malysh is a tiny blowback-operated pistol that is obviously intended for deep concealment and undercover work or as a backup weapon. It is so small that its magazine holds only five rounds of 9x18mm ammunition. The rudimentary sight is no more than a groove in the top of the slide, similar to the Smith & Wesson Sigma 380. Unlike the Sigma, the OTs-21 appears to be constructed entirely of metal, although the exact type is not clear. Photographs have shown two distinct colors on the frame and slide of the pistol, but there have been no indications as to its design and construction.

OTs-21 *MALYSH* SPECIFICATIONS

Caliber	9x18mm
Operation	Blowback, Semiautomatic
Length	4.9 in. (126mm)
Weight	11.3 oz. (325 gr.)
Barrel Length	2.5 in. (63.5mm)
Effective Range	15–20 m.
Magazine Capacity	5 Rounds
Sights	Groove

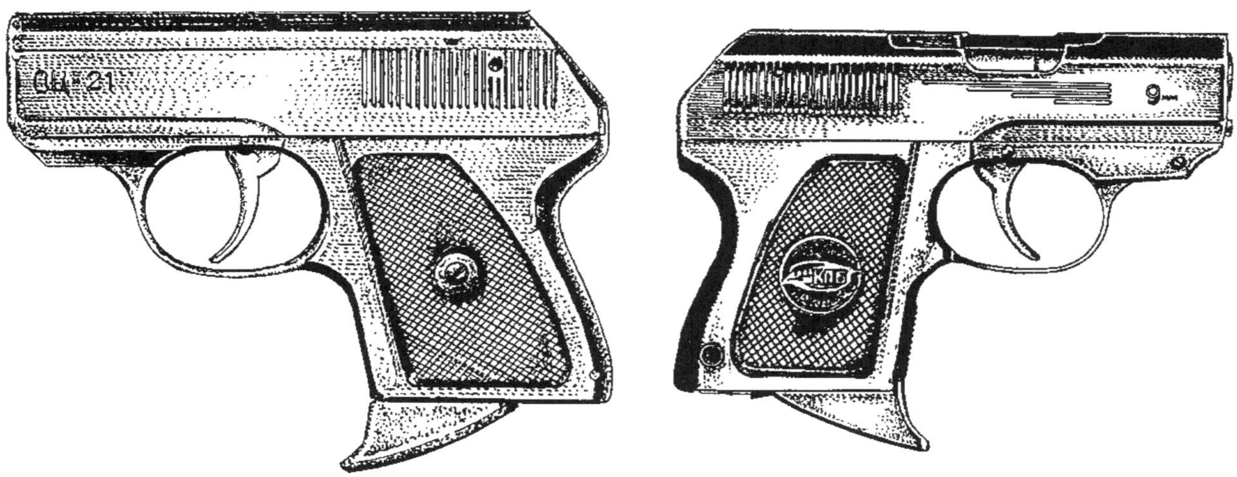

Left and right view of the OTs-21 Malysh *automatic pistol in 9x18mm with a five-shot magazine.*

OTs-27 *BERDYSH* (POLEAX) PISTOL

The OTs-27 is another unusual pistol manunfactured by Tula KBP. The original design was intended to be a multicaliber firearm that could, by changing between two barrels and magazines, fire 9x18mm, 9x18mmM, 9x19mm, or 7.62x25mm ammunition. The barrel and magazine change was claimed to be necessary to adapt the pistol for the 7.62x25mm cartridge. The fielded version of the OTs-27, however, is chambered only for both standard and high-velocity 9x18mm cartridges. By changing barrels and magazines, 9x19mm can be fired from the pistol. Why this became necessary is unknown, because the forces generated by the 9x18mmM cartridge are very similar to those of the 9x19mm. Despite Russian claims that the OTs-27 operates by blowback, the pistol probably incorporates delayed-blowback operation, since blowback operation would require a very heavy and unwieldy slide and recoil spring. If the Berdysh does use the delayed-blowback principle, it probably is a method similar to that used in the PMM (i.e., grooves in the chamber to delay extraction sufficiently for pressures to drop and keep the slide velocity down when firing the "hotter" cartridges). Like other recent Russian pistols, the OTs-27 has fittings for mounting a laser designator and suppressor. Tula KBP claims that the pistol is in service with the MVD.

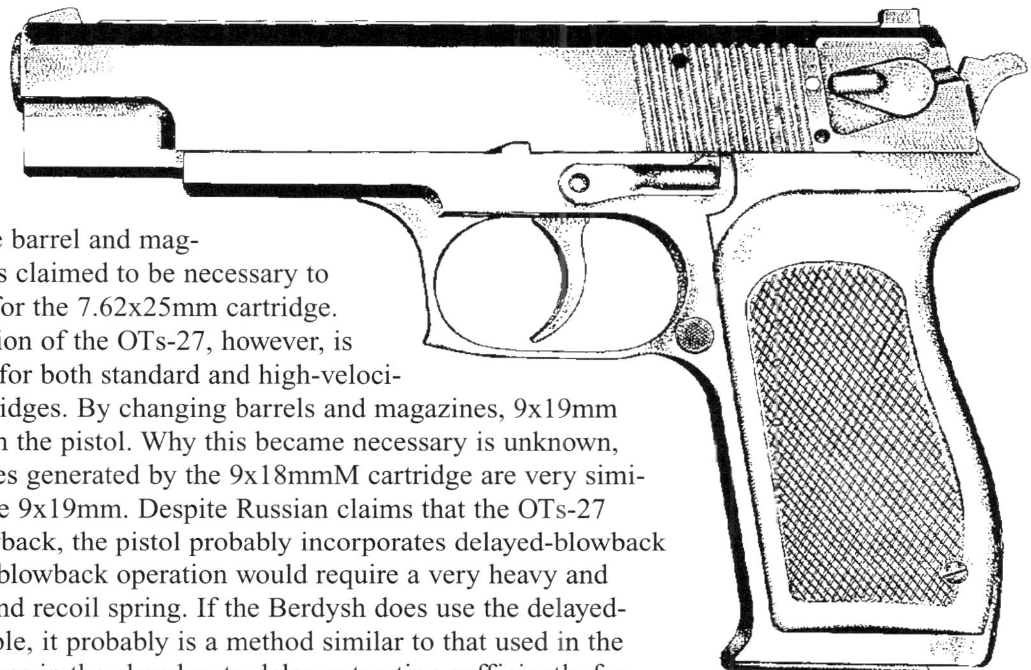

The OTs-27 Berdysh *automatic pistol.*

OTs-27 *BERDYSH* PISTOL SPECIFICATIONS

Caliber	9x18mm/9x18mmM/9x19mm
Operation	Delayed Blowback, Semiautomatic
Weight	34 oz. (.96 kg.)
Length	7.8 in. (200mm)
Barrel Length	App. 5 in. (127mm)
Effective Range	25 m.
Sights	Front, Blade; Rear, Notch, Adjustable for Windage

5.45x18mm OTs-23 *DROTIK* (JAVELIN) PISTOL

The OTs-23 *Drotik* is a select-fire pistol manufactured by Tula KBP. The overall concept is similar to the 1950s-era Stechkin, which is still in production in a semiautomatic-only version, according to Tula KBP sources. There is an obvious compensator slot atop the barrel at the muzzle of the pistol that helps prevent muzzle rise when firing in burst mode. This, in combination with the burst-fire feature and the 5.45x18mm caliber, probably goes far toward making the OTs-23 controllable on burst fire. The diminutive 5.45x18mm round is a good candidate for use in a select-fire pistol because of its low recoil impulse and ability to penetrate body armor at typical CQB ranges. Rather than firing in full automatic, however, the OTs-23 fires in three-round bursts in full-automatic mode. In burst mode, the OTs-23 fires at a cyclic rate of 1,700 RPM. Given such a high cyclic rate, the three-round burst leaves the barrel almost before the pistol has had a chance to recoil, thereby increasing hit probability. This is a virtual necessity given the ballistics of the 5.45x18mm cartridge.

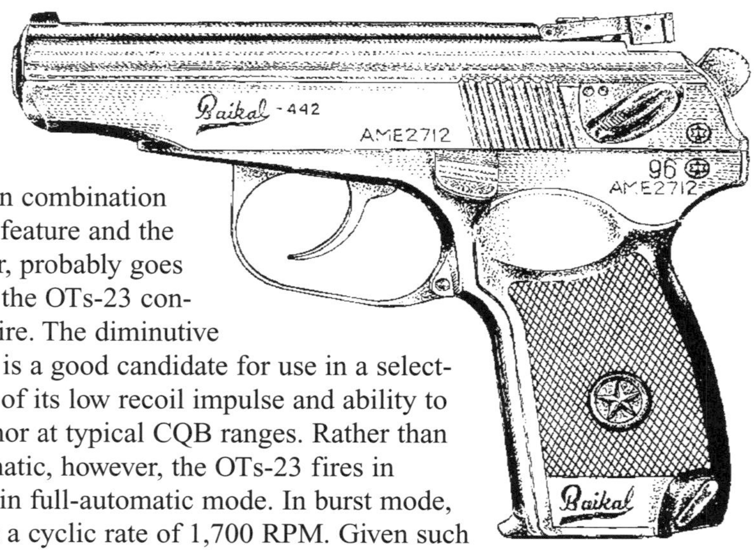

The OTs-23 Drotik *automatic pistol in 5.45x18mm.*

The *Drotik* is unusual for a Russian pistol in that it appears to be derived at least in part from Beretta designs. The pistol has an open slide similar to that on Beretta pistols and what appears to be a Beretta-style takedown lever-release button above the trigger guard. Like many current tactical pistols, the *Drotik* has grooves on the forward lower portion of the frame for fitting laser aiming devices or white-light accessories. According to Tula sources, the OTs-23 *Drotik* is in service with the MVD troops.

OTs-23 *DROTIK* SPECIFICATIONS

Caliber	5.45x18mm
Operation	Blowback, Select Fire
Weight	33.8 oz. (.96 kg.)
Length	7.6 in. (195mm)
Barrel Length	App. 5 in. (127mm)
Effective Range	25 m.
Sights	Front, Blade; Rear, Notch, Adjustable

OTs-33 *PERNACH* PISTOL

The OTs-33 is very similar to the OTs-23, previously discussed, save for the different caliber and slide configuration. The slide of the OTs-33 is closed with an ejection port, rather than open as with the OTs-23. According to Russian sources at Tula KBP, the OTs-33 is intended to be a replacement for the APS Stechkin, which was still in at least limited production in a semiautomatic version as of early 1998. The OTs-33 is especially curious in that Tula claims it to be fully automatic, whereas the OTs-23 is burst fire only. The Stechkin was only marginally controllable on full auto and even at a cyclic rate of 750 RPM emptied its 20-round magazine so quickly that a reload was necessary in very short order. Nonetheless, some agency in the MVD apparently believed that the Stechkin was a good idea and wanted a replacement that was even better. The *Pernach* fits that requirement to a T. It has a larger magazine capacity than the Stechkin and fires at a slightly higher cyclic rate. The new pistol also fires both standard and high-velocity 9x18mm cartridges; full automatic fire with the latter, especially without the benefit of a compensator such as that on the OTs-23, should provide a real challenge to the shooting abilities of the MVD troops who use it, even with the oddly shaped folding metal shoulder stock installed! Whether the OTs-33 will prove to be more useful in service than the Stechkin remains to be seen, but the new pistol is reportedly now in service with the Russian MVD troops. Like virtually all new Russian pistols, the OTs-33 has a provision for a laser designator and a suppressor.

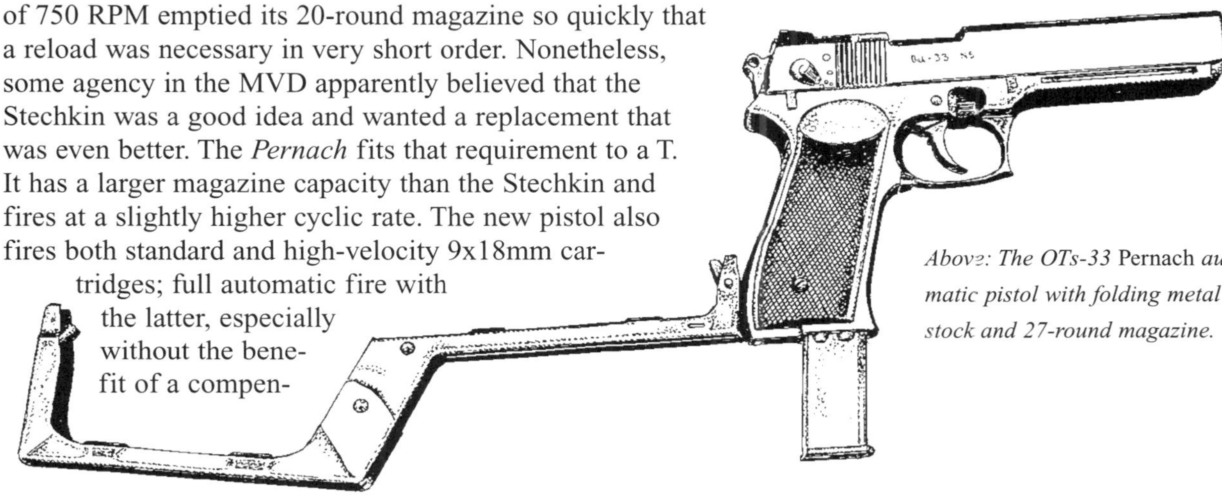

Above: The OTs-33 Pernach *automatic pistol with folding metal stock and 27-round magazine.*

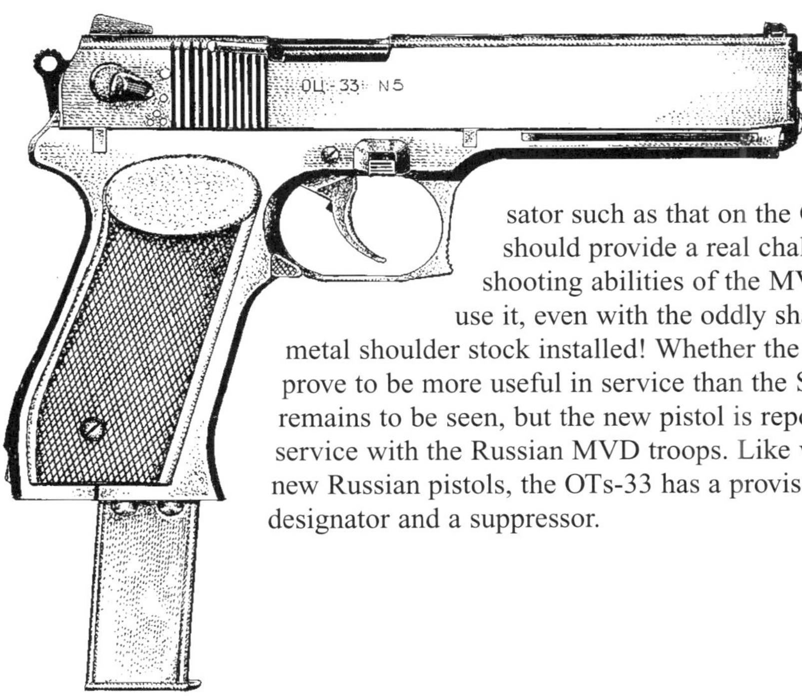

Left: The OTs-33 Pernach *shown with selector on semiautomatic with 27-round magazine.*

OTs-33 *PERNACH* SPECIFICATIONS

Caliber	9x18mm/9x18mmM
Operation	Delayed Blowback, Select Fire
Weight	41.9 oz. (1.15 kg.)
Length, Without Stock	8.7 in. (221mm)
With Stock	20.9 in. (531mm)
Barrel Length	App. 5.5 in. (140mm)
Effective Range	25 m.
Magazine Capacity	18 or 27 Rounds
Cyclic Rate	800 RPM
Sights	Front, Blade; Rear, Notch, Adjustable

5.45x18mm *DREL* (DRILL) PISTOL

The *Drel* pistol is derived from the .22LR *Margo* pistol that has been sold for sporting use in Russia for a number of years. The Drel was designed to an MVD and militia (police) requirement for a compact pistol. Like the other 5.45x18mm pistols developed by Russian manufacturers, the *Drel* is a straight blowback design. The number and variety of compact pistols developed for the 5.45x18mm cartridge indicate that the Russians are very satisfied with the cartridge's performance within the bounds of its design and intended use.

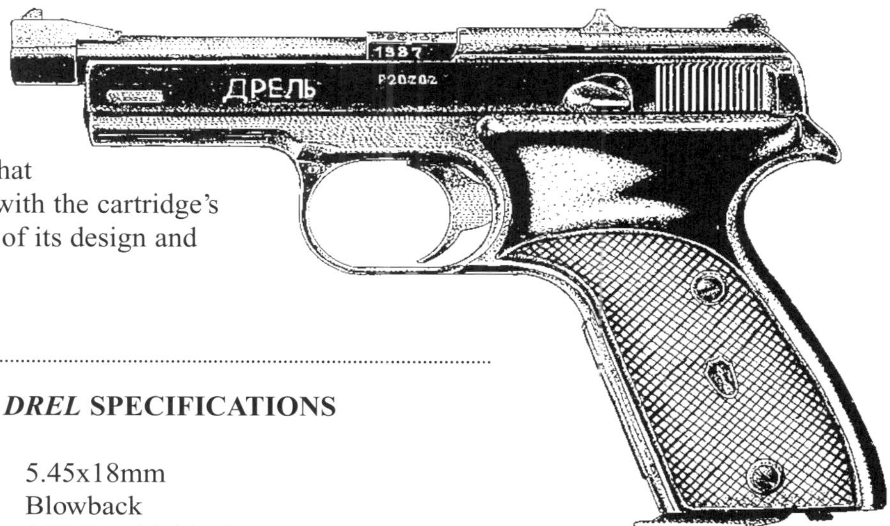

The Drel *5.45x18mm automatic pistol with 10-round magazine.*

5.45x18mm *DREL* SPECIFICATIONS

Caliber	5.45x18mm
Operation	Blowback
Weight	1.23 lbs. (.56 kg.)
Length	7.4 in. (190mm)
Barrel Length	App. 4 in. (101mm)
Effective Range	25 m.
Magazine Capacity	10 Rounds
Sights	Front, Blade; Rear, Notch

P-96 9x19mm PISTOL

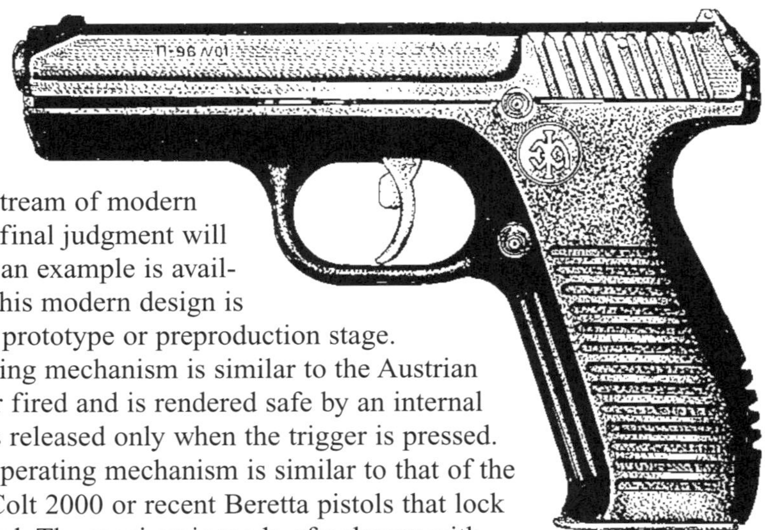

The P-96 plastic-frame pistol.

The Tula P-96 pistol represents the first of a new generation of pistols for Russia. Unlike the *Gurza*, which is chambered for an unusual cartridge and whose functioning mechanism is basically unsafe as currently maunfactured, the P-96 appears to be in the mainstream of modern pistol design, but final judgment will have to wait until an example is available to test fire. This modern design is apparently still in prototype or preproduction stage.

The P-96 firing mechanism is similar to the Austrian Glock; it is striker fired and is rendered safe by an internal mechanism that is released only when the trigger is pressed. The short-recoil operating mechanism is similar to that of the Stoner-designed Colt 2000 or recent Beretta pistols that lock via a rotating barrel. The receiver is made of polymer with molded-in metal rails on which the slide rides, and the magazine release is located so that it can be readily released by the shooter's left thumb. Significantly, the P-96 is chambered only in 9x19mm.

Examples of the P-96 I have seen in photographs have appeared well-finished with a matte black polymer receiver and a stainless or electroless nickel-finished slide. All surfaces are rounded with no sharp edges or protrusions to snag on clothing. The P-96 appears to be the first Russian firearm designed from the outset to compete in the international market, rather than an adaptation of an existing military or police firearm. It is also notable in that it was designed for the 9x19mm cartridge from the outset and is not even offered in 9x18mm.

P-96 9x19mm PISTOL SPECIFICATIONS

Caliber	9x19mm Parabellum
Operation	Short Recoil, Semiautomatic
Weight	20.1 oz. (.57 kg.)
Length	7.4 in. (188mm)
Barrel Length	App. 5 in. (127mm)
Effective Range	25 m.
Magazine Capacity	18 Rounds
Sights	Front, Blade; Rear, Notch, Adjustable for Windage

9x18mm OTs-01 *KOBALT*/REVOLVER STECHKINA AVRAMOVA (RSA)

The Russians have never been known for conventional revolver designs, but the OTs-01 *Kobalt/RSA* comes closest to anything that their arms industry has ever produced that can be described as a "conventional" revolver. In its prototype form, the revolver was known as the RSA, but it is now listed as the OTs-01 *Kobalt* and is reportedly in service with the Russian MVD. The *Kobalt* is available in two versions, the OTs-01 in 9x18mm caliber and the OTs-01S in 9x17mm (.380 ACP). Both are loaded via oddly shaped, three-round clips. It has been advertised in Russian international sales documents, but this revolver and those that follow do not appear to be truly competitive with such Western revolvers as Colt, Smith & Wesson, Taurus, and Rossi.

The OTs-01 appears to be loosely based on Smith & Wesson designs, but with differences. From what can be seen in available photographs, the metal in the frame is roughly finished and appears to be made of cast alloy rather than forged steel. This conclusion is also supported by the fact that the OTs-01 apparently has a barrel liner, which would not be necessary were the barrel of forged steel. The cylinder and internal components apparently are made of steel rather than alloy. The type of construction employed in the OTs-01 is typical of inexpensive European revolvers, which in no way detracts from their intended purpose. Such revolvers are simply not up to the quality standards of Western designs and thus cannot be expected to meet the longevity or reliability standards of the Western models. Service life of the OTs-01 is listed by Russian sources as only 3,000 rounds, a fraction of what can be expected of a Western revolver and yet another indicator that the frame is cast alloy.

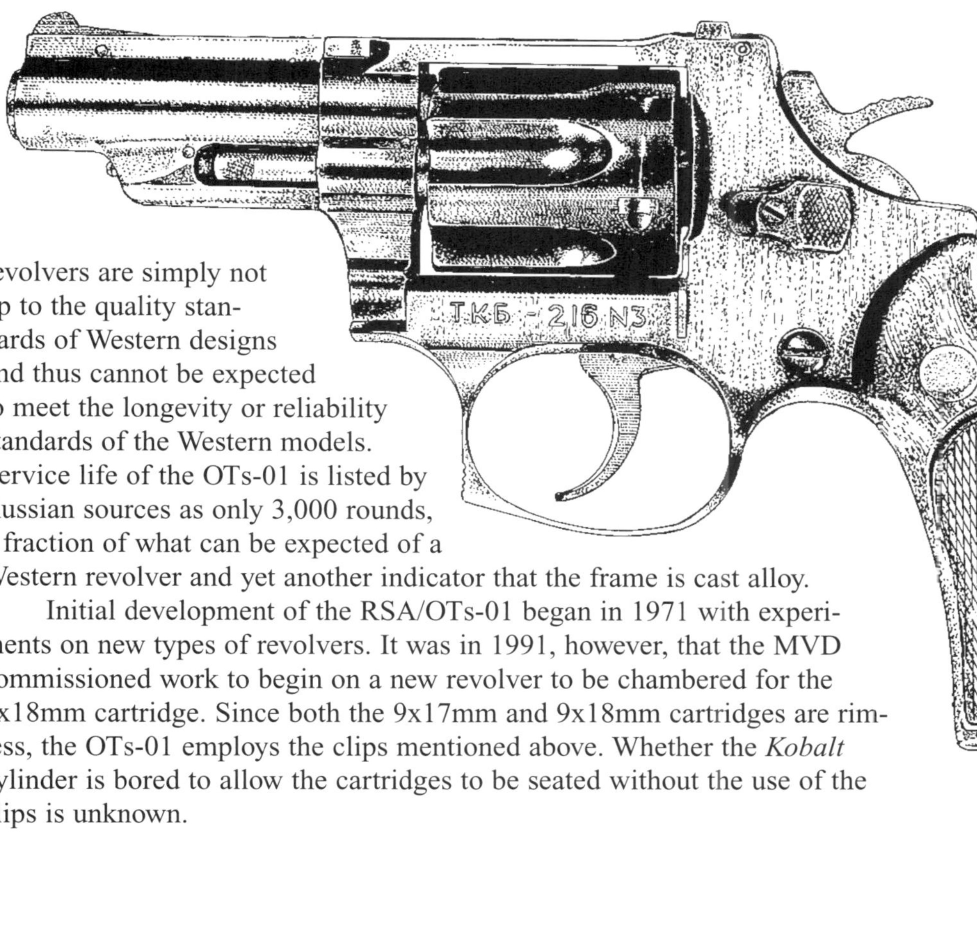

The six-shot RSA 9mm revolver in 9mm PM.

Initial development of the RSA/OTs-01 began in 1971 with experiments on new types of revolvers. It was in 1991, however, that the MVD commissioned work to begin on a new revolver to be chambered for the 9x18mm cartridge. Since both the 9x17mm and 9x18mm cartridges are rimless, the OTs-01 employs the clips mentioned above. Whether the *Kobalt* cylinder is bored to allow the cartridges to be seated without the use of the clips is unknown.

The OTs-01 and OTs-015 Kobalt *revolvers were developments of RSA. The OTs-01 was in 9x18mm PM, and the OTs-015 was in 9x17mm Kurz (.380 ACP).*

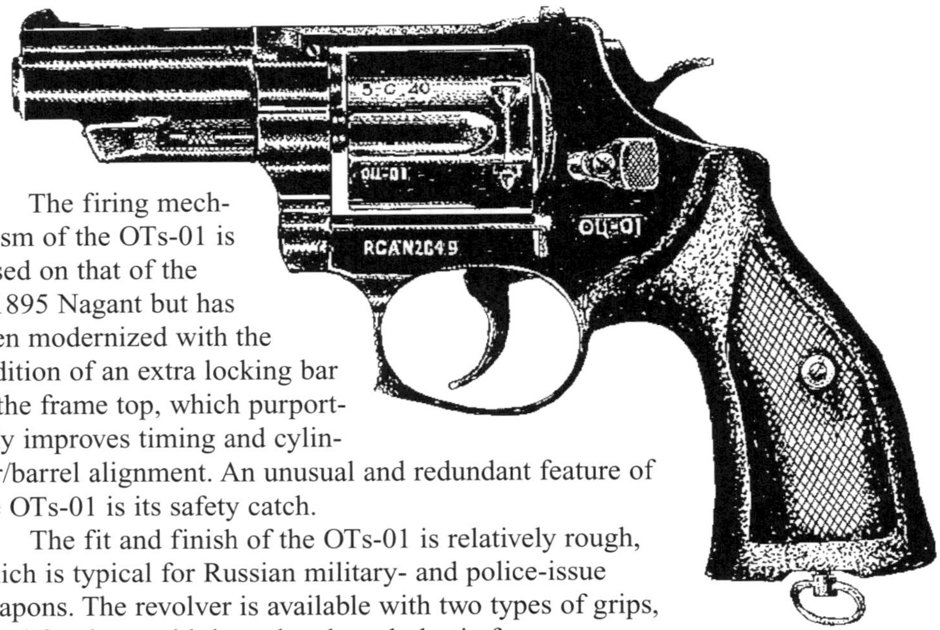

The firing mechanism of the OTs-01 is based on that of the M1895 Nagant but has been modernized with the addition of an extra locking bar in the frame top, which purportedly improves timing and cylinder/barrel alignment. An unusual and redundant feature of the OTs-01 is its safety catch.

The fit and finish of the OTs-01 is relatively rough, which is typical for Russian military- and police-issue weapons. The revolver is available with two types of grips, wood for those with large hands and plastic for persons with small hands. According to sources at Tula KBP, the OTs-01 is in service with Russian MVD forces.

OTs-01/RSA REVOLVER SPECIFICATIONS

Caliber	9x17mm/9x18mm
Operation	Double-/Single-Action Revolver
Weight	28.2 oz. (.8 kg.)
Length	7.8 in. (157mm)
Barrel Length	App. 4 in. (101mm)
Effective Range	25 m.
Cylinder Capacity	6 Rounds
Sights	Front, Blade; Rear, Notch, Adjustable for Windage

9x18mm R-92/R-92KS REVOLVER

Like the OTs-01/RSA revolver described above, the R-92 is stated by the Russians to be for use against "hostile manpower." Both appear in the *Russian Military Catalog* and thus are available for sale on the international arms market, but in its own way the R-92 is even less competitive than the OTs-01/RSA when compared with such firearms as those manufactured by Colt and Smith & Wesson. Both the R-92 and the OTs-01 are manufactured by Tula KBP, and both are claimed to be in service with the Russian MVD. Like the OTs-01, the R-92 is available in 9x17mm (.380 ACP), obviously an effort to make it more attractive in Western markets.

The R-92 is a double-action-only, 5-shot revolver that loads from a "full-moon" clip that contains all five cartridges. The R-92 cylinder will accept cartridges without the clip, but they cannot be simultaneously ejected and must be picked out by hand one at a time. The frame of the R-92, like that of the OTs-01, appears to be cast from nonferrous alloy with a steel cylinder and pressed-in-steel barrel. Sights are simple, with a nonadjustable rear notch cast in the top strap of the frame and a blade front sight. The R-92 is expressly designed for concealed carry, and although its overall configuration (with "forward displaced" trigger and grip) lends to this capability, the overall design of the R-92 gives it an appearance that can at best be described as awkward.

Although I have not fired this revolver, the relationship of the grip to the trigger and barrel is such that it is difficult to envision good ergonomics and pointing abilities for the weapon. Moreover, the barrel sits far above the hand of the shooter, which results in increased felt recoil and muzzle rise. In a low-impulse revolver, such as the R-92, this may not be a great disadvantage, but since it is a revolver, the R-92 can be expected to accept and fire the new high-velocity 9x18mm cartridges. In addition, one variant of the much larger *Udar* revolver (see p. 77) appears to be no more than a scaled-up version of the R-92. There is also a gas version of

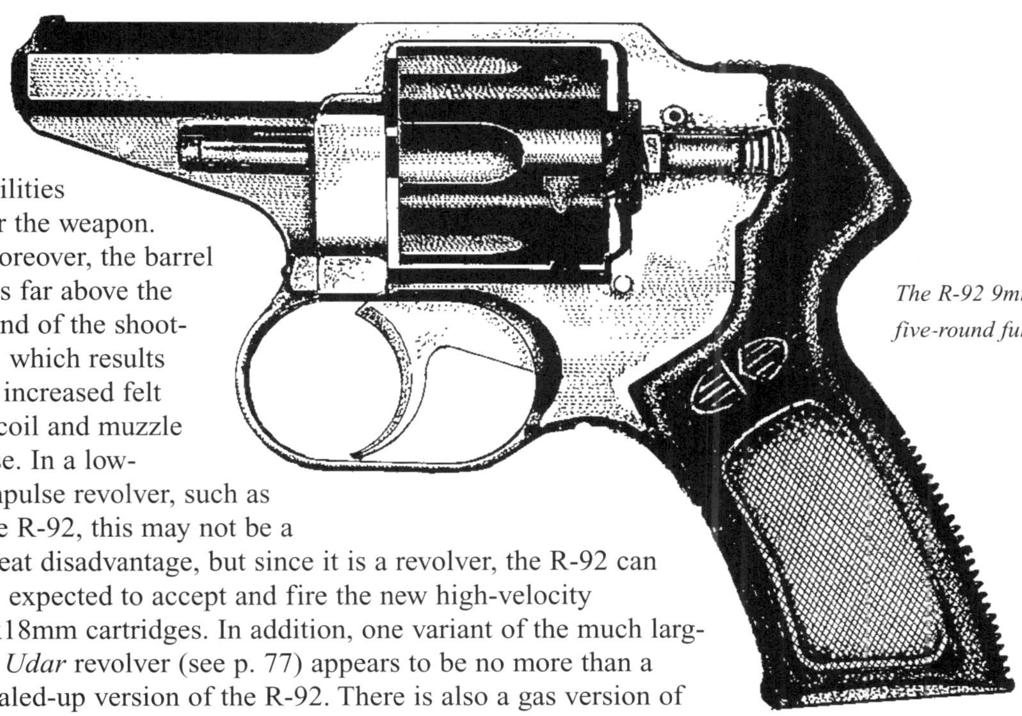

The R-92 9mm PM revolver with five-round full-moon clip.

the R-92 that cannot fire live ammunition, but instead fires only irritant-gas cartridges. This version is designated the GP-92.

R92/R-92KS REVOLVER SPECIFICATIONS

Caliber	9x18mm/9x17mm
Operation	Revolver, Double Action Only
Weight	18.3 oz. (.52 kg.)
Barrel Length	App. 2.5 in. (63mm)
Effective Range	25 m.
Cylinder Capacity	5 Rounds
Sights	Front, Blade; Rear, Notch

12.3mm *UDAR* (BLOW) LARGE-CALIBER REVOLVER (VERSIONS 1 AND 2)

As suggested by the heading of this section, there are two revolvers that carry the Russian nickname *Udar*, which means a blow in the sense of striking someone with a hammer or ball bat rather than other connotations that might come to a more deviant mind. Both versions of the *Udar* appear to be "scaled-up" versions of the OTs-01 revolver and R92 revolvers, respectively. The existence of two such revolvers would cause little comment in the West, but in Russia, where ownership of pistols is restricted to the military, police, and criminals, and where most firearms are designed around a government requirement, these revolvers are perplexing.

It is possible that they were competitors for a government contract, and one lost and is now being offered on the international arms market. If that is the case, the R-92-derived *Udar* may have been the loser, as it has been shown at numerous arms shows and has been featured in *Military Parade*, the official trade magazine of the Russian military, which features articles on hardware that the Russian government hopes to sell, from small arms to submarines. On the other hand, the OTs-01-derived *Udar* is prominently featured in the *Russian Military Catalog*, which is ostensibly the definitive list of Russian weapons and military equipment available for export, while the R-92-derived Udar is absent. It is also possible that both Udars were accepted into service. At this point, I will leave the mystery of the two Udars and proceed to address their individual characteristics.

Udar Version 1

This version of *Udar* was developed by Tula KBP to a Russian MVD requirement. The result was a very large caliber revolver that fires a number of different types of 12.3x42Rmm ammunition and is available in three different versions, *Udar-S*, *Udar-T*, and *Udar*. The Udar-S (Service) fires a special 12.3x22mm "service" cartridge with a lighter bullet and lower muzzle velocity than the standard Udar and apparently is available to individual citizens with "proper licenses," if the Russian material pertaining to the revolver is correct. The *Udar-T* (Training) fires a only a paint bullet and nonlethal rubber bullet. The Udar fires a variety of 12.3x40mm cartridges, which are fully discussed in the ammunition chapter of this book. The full-power cartridges are most likely 12.3x50Rmm, while the lower power versions for the *Udar-S* and *Udar-T* revolvers are 12.3x46Rmm

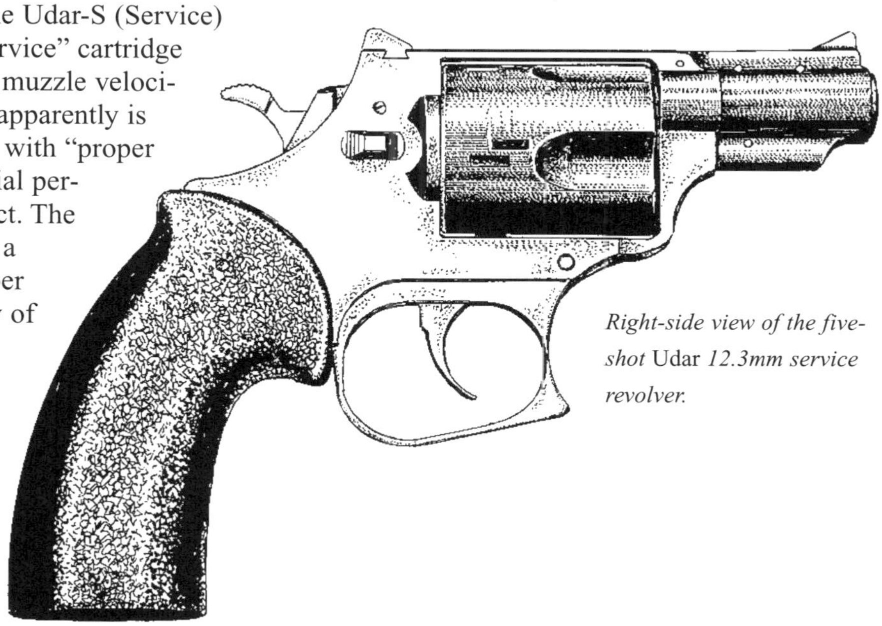

Right-side view of the five-shot Udar *12.3mm service revolver.*

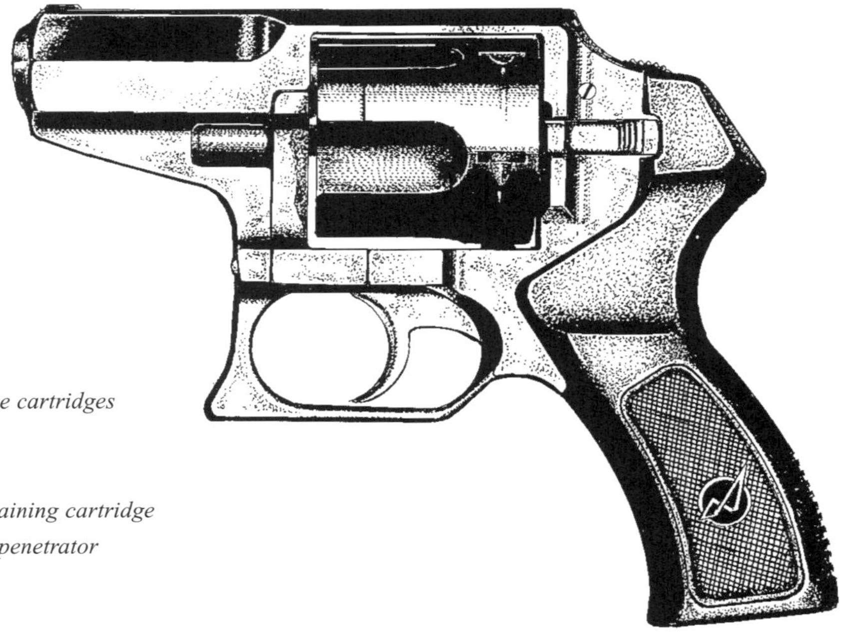

Left-side view of the five-shot Udar *12.3mm service revolver.*

The Udar-S *large-caliber revolver. The cartridges shown are as follows:*
Top: Steel bullet, 8-shot lead
Middle: Paint bullet, rubber bullet, training cartridge
Bottom: Steel bullet, sabot with steel penetrator

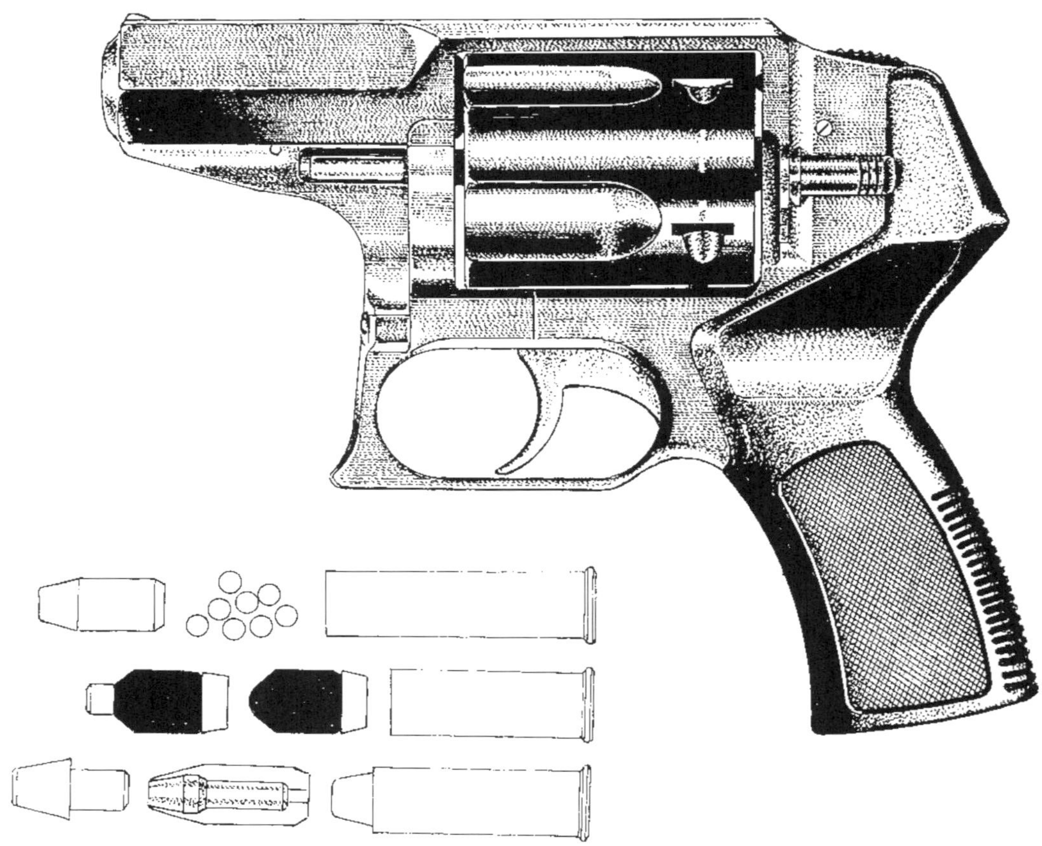

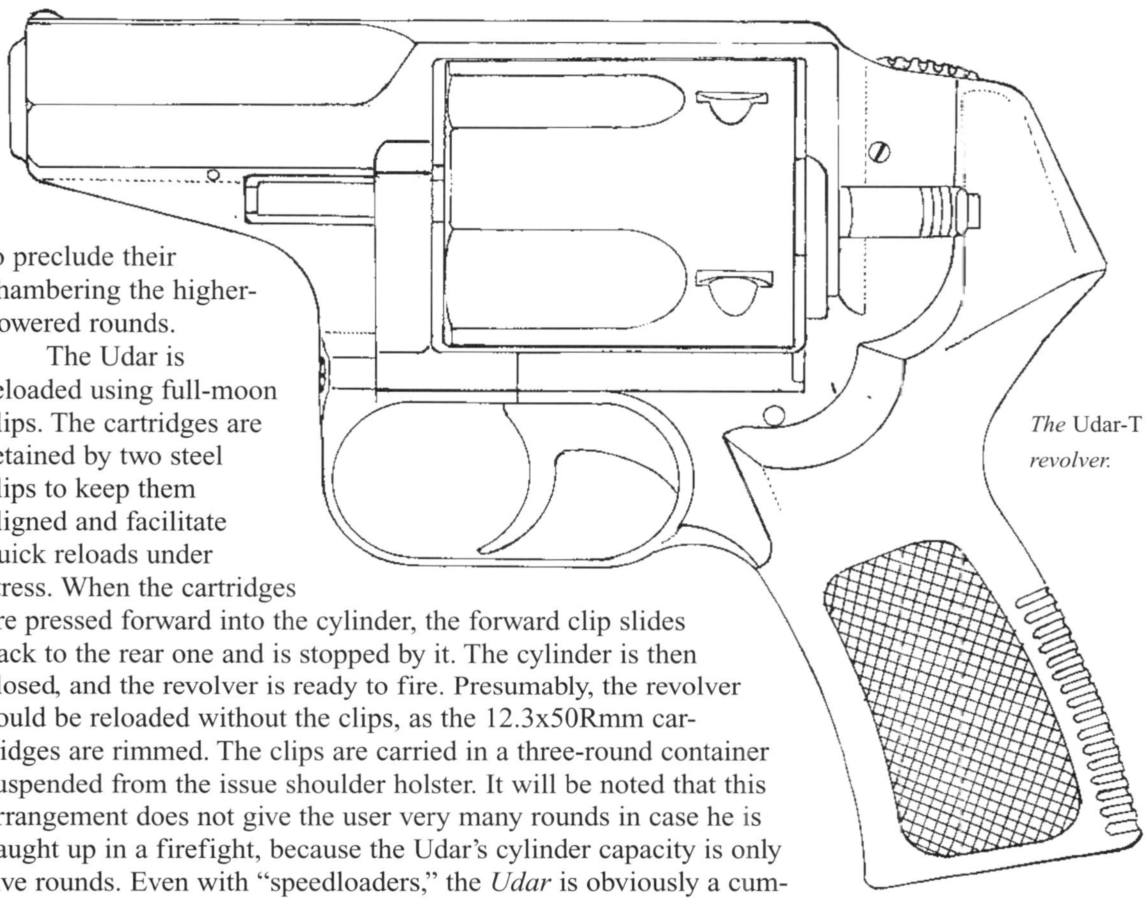

The Udar-T *training revolver.*

to preclude their chambering the higher-powered rounds.

The Udar is reloaded using full-moon clips. The cartridges are retained by two steel clips to keep them aligned and facilitate quick reloads under stress. When the cartridges are pressed forward into the cylinder, the forward clip slides back to the rear one and is stopped by it. The cylinder is then closed, and the revolver is ready to fire. Presumably, the revolver could be reloaded without the clips, as the 12.3x50Rmm cartridges are rimmed. The clips are carried in a three-round container suspended from the issue shoulder holster. It will be noted that this arrangement does not give the user very many rounds in case he is caught up in a firefight, because the Udar's cylinder capacity is only five rounds. Even with "speedloaders," the *Udar* is obviously a cumbersome weapon to reload under the best of conditions.

Udar Version 2

Less is known of this version of the *Udar* because information on it has not been as widely publicized by the Russians. It is manufactured by TsNIITochmash and is clearly a more conventional firearm design. This *Udar* is chambered in 12.3x50Rmm cartridge. Why the Russians would manufacture two different calibers for two revolvers is a question that only the Russians themselves can answer. Presumably, this version of *Udar* can accept the 12.3x40mm and 12.3x22mm cartridges of the Tula-manufactured weapon. Like version 1, version 2 holds five rounds, but there is only one type of this variant of *Udar*, and according to Russian sources, it fires ball, armor-piercing, plastic bullet, gas, and noise rounds interchangeably. This version of *Udar* was also developed for use by "law and order enforcement agencies against people violating public order." Like the version 1 *Udar*, version 2 is reloaded from a five-round clip. Whether clips interchange between *Udars* is unknown, and although logic dictates that clips would be interchangeable between the two *Udar* revolvers, especially when both were apparently designed for law enforcement use, this might well not be the case.

UDAR (VERSION 1) SPECIFICATIONS

Caliber	12.3x40Rmm (*Udar*) 12.3x40Rmm (*Udar-S*, *-T*)
Operation	Revolver, Double or Single Action
Weight	32.4 oz. (.92 kg.)
Length	6.8 in. (173mm)
Barrel Length	App. 2.5 in. (63.5mm)
Effective Range	App. 50 m.
Cylinder Capacity	5 Rounds
Sights	Front, Blade; Rear, Groove in Frame Top

UDAR (VERSION 2) SPECIFICATIONS

Caliber	12.3x50Rmm; 12.3x46Rmm; 12.3x40Rmm (?); 12.3x22Rmm (?) 12.3x40Rmm (*Udar-S, -T*)
Operation	Revolver, Double or Single Action
Weight	33.5 oz. (.95 kg.)
Length	8.2 in. (210mm)
Barrel Length	App. 2.5 in. (63.5mm)
Effective Range	App. 50 m.
Cylinder Capacity	5 Rounds
Sights	Front, Blade; Rear, Notch, Adjustable for Windage

OTs-20 *GNOM* (GNOME) 12.5x40Rmm REVOLVER

To add to the questions surrounding the two Udar revolvers, the OTs-20 *Gnom* recently was announced by Tula KBP, which produces the version 1 Udar. The capabilities of the OTs-20 revolver are so similar to those of the Udar that it is difficult to imagine why Tula produces two guns that are so alike, especially when both are stated to be in service with the Russian MVD. There are, however, differences between the two seemingly similar revolvers.

The OTs-20 is a smoothbore and thus is intended for short-range work. The OTs-20 fires three types of cartridges: one with a steel bullet that penetrates 4.5mm of (type unspecified) steel armor plate at 25 meters, another with a lead bullet for "high stopping power," and a shot shell for "use when there is no information on the precise location of the enemy." The *Gnom* cartridges are very similar to those used in the *Udar*. In fact, the *Udar* and OTs-20 cartridges are both stated to have been derived from 32-gauge hunting cartridges, although those for the *Udar* are stated to be 12.3mm in diameter, while those for the *Gnom* are claimed to be 12.5mm. It will be recalled that the 12.3mm cartridges for the *Udar-S* and *Udar-T* are 12.3x22Rmm, while those for the standard revolver are 12.3x40Rmm. The cartridges for the OTs-20 are claimed to be 12.5x40Rmm. The slightly larger diameter of the *Gnom* cartridges may be intended to prevent them from being loaded into *Udar-S* or *Udar-T* revolvers. It is difficult, however, to imagine that the steel and lead bullets fired from the OTs-20 would be very accurate from the smoothbore revolver, especially given their slightly smaller diameter, unless there were some special design features in the bullets themselves.

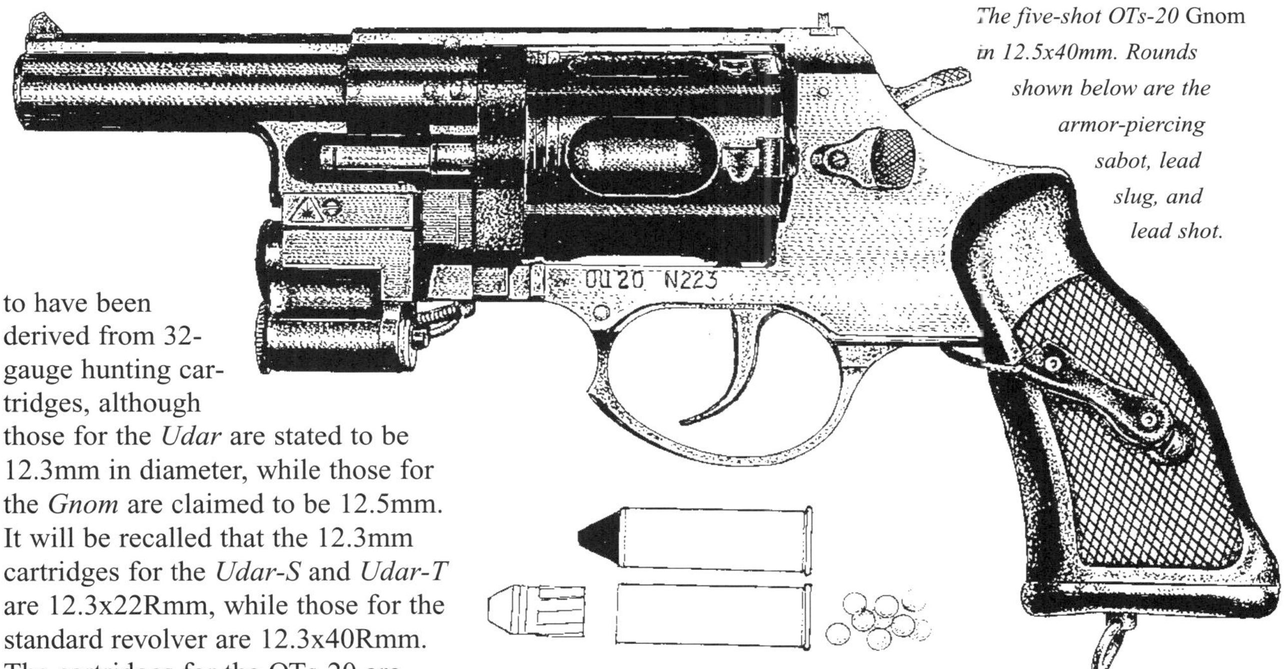

The five-shot OTs-20 Gnom *in 12.5x40mm. Rounds shown below are the armor-piercing sabot, lead slug, and lead shot.*

OTs-20 *GNOM* REVOLVER SPECIFICATIONS

Caliber	12.5x40Rmm
Operation	Revolver, Double or Single Action
Weight	39 oz. (1.1 kg.)
Length	9.8 in. (250mm)
Barrel Length	App. 5.5 in. (140mm)
Effective Range	25 m.
Cylinder Capacity	5 Rounds
Sights	Front, Blade; Rear, Notch, Adjustable for Windage

4.5x39mm SPP-1/SPP-1M UNDERWATER PISTOL

Along with the APS underwater assault rifle, the SPP-1/SPP-1M underwater pistol was developed in the late 1960s for use by Soviet SPETsNAZ forces. The SPP-1 was designed by Vladimir Simonov, who is justly famous for also designing the SKS carbine. The SPP-1 was adopted for service in 1971 but remained completely unknown in the West until after the demise of the Soviet Union, when its manufacturer, TsNIITochmash, began advertising it for sale and export.

The SPP-1 and SPP-1M are essentially identical, save for the size of the trigger guard, which has been enlarged on the SPP-1M to accommodate the bulky gloves worn by divers. The SPP-1 operation is quite simple. The firing mechanism is double action only and incorporates a rotating firing pin much like that of the 19th-century American Sharps derringer. In fact, the SPP-1 could be considered a modified and enlarged version of this antique design. Nonetheless, the SPP-1 is reliable and effective, which is what is necessary in a military firearm. The SPP-1 is a simple tip-open, four-barreled pistol. There is no ejection per se; the four-round clip is extracted and then manually removed from the pistol. This is not necessarily a shortcoming; in clandestine operations, one may not want to leave discarded cartridge casings lying about, even under water, so manual extraction may be by intent. Reloading is via a four-round clip that is simply inserted into the barrel breeches. To accomplish this, the safety is set into the *zaryazh* (load) position, which releases the barrel lock, allowing the barrel group to be tipped open for reloading. Once the barrel group is closed and the external safety set to *ogon* (fire), the SPP-1 is ready to fire.

In the safe position, the safety blocks the firing mechanism and firing spring. When the trigger is pulled, the connecting lever visible in the illustration moves the firing pin to the rear and compresses the firing spring. As the trigger continues to the rear, it engages a groove, which rotates the firing mechanism 90 degrees clock-

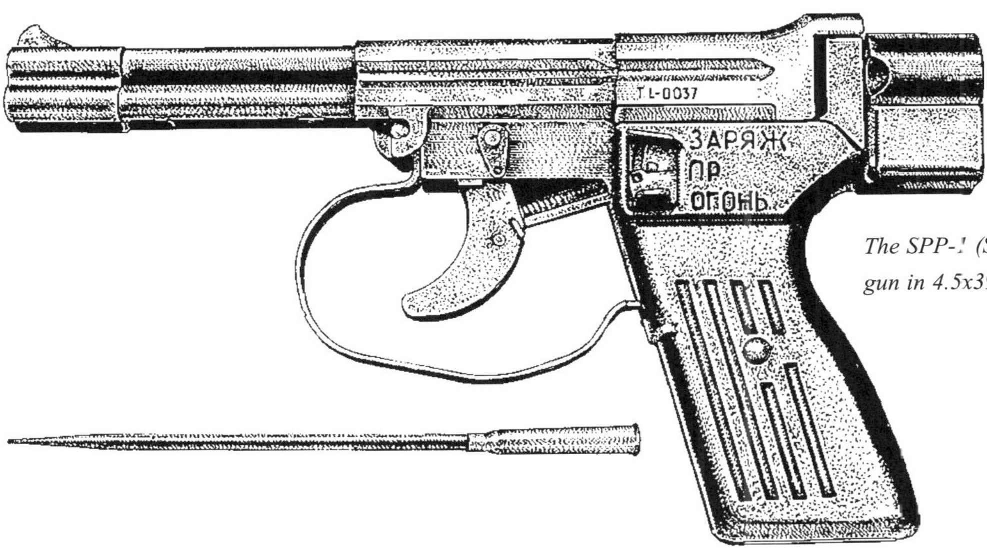

The SPP-1 (SPP-1M) multibarrel handgun in 4.5x39mm.

wise. As the mechanism comes into alignment with the barrel to be fired, it locks into place and the sear is released, firing the pistol.

Like the APS rifle, the SPP-1 pistol fires a steel dart, which is stabilized in water by the flow of water along its surfaces rather than by rotation. The SPP-1 is thus also a smoothbore. The SPS 4.5x39mm cartridge has an effective range from 11 to 17 meters, depending on depth. The bullet does not stabilize in air. Details of the SPS cartridge may be found in the ammunition chapter of this book. The SPP-1's "basic load" issued to combat divers is three sealed reloading packets, each with one clip of four rounds (*Russian Military Catalog*, Part 1 [IV], p.105).

SPP-1/1M 4.5x30Rmm UNDERWATER PISTOL SPECIFICATIONS

Caliber	4.5x39mm
Operation	Manual, Rotating Firing Pin
Weight	2.09 lbs. (.95 kg.)
Length	9.6 in. (244mm)
Barrel Length	App. 7 in. (178mm)
Effective Range	11–17 m., Depending on the Depth
Cylinder Capacity	4 Rounds
Sights	Front, Blade; Rear, Notch

CHAPTER THREE

SUBMACHINE GUNS

The Russians have had a love affair with submachine guns dating back to World War II (in Russia, the Great Patriotic War). Indeed, they fielded literally millions of PPSh-41 and PPS-43 submachine guns and used them to great effect against the German invaders. A lasting image of the Eastern Front is that of a Russian soldier charging the hated Germans with his PPSh-41 in hand. With the coming of the AK-47, which combined the best features of the submachine gun and rifle, submachine guns were abandoned, other than a few experimental models designed in the late 1940s. As can best be determined, the last of these was designed by Simonov in 1949, but this model never got beyond the prototype stage.

There was brief renewed interest in submachine guns in the late 1960s when the Soviets became aware of Western designs, such as the Ingram, and they considered such weapons for special operations because of their ability to be concealed while maintaining a fully-automatic firepower capability. There was some work done by, among others, Dragunov, but the Soviet military considered bullet dispersion of the experimental weapons beyond 50 meters unacceptable, so efforts were dropped.

By the late 1980s, however, the Soviet Union was falling apart, and along with it, law and order. Russian police requested a new submachine gun and in 1991–92 the arms industry responded by resurrecting the best of the 1960s designs—Dragunov's. This weapon was modified, modernized, refined, and fielded in 1993 as the *Kedr* (Konstruktsiya Evgeniya Dragunov; also meaning *cedar* in Russian). The *Kedr* was later developed into the more powerful *Klin*. Both weapons will be discussed below. *Kedr* and *Klin* production apparently has been insufficient to meet Russian law enforcement demands, or perhaps other designs improve on their performance. For whatever reasons, there has been a renaissance in Russian submachine guns, as we shall presently see by the number of these weapons now being offered for sale both internally and on the international arms market.

Unlike designs of the Great Patriotic War era, the new Russian submachine guns are generally very compact and almost all can be optionally equipped with suppressors.

It must be noted that the status of many of the submachine guns that are covered here is unclear. Most have been shown at various Russian arms shows, and some have actually been fired by Westerners, but it is also clear that some are prototypes that may never see full production. I will attempt to differentiate between those weapons that are actually in production and those that exist only in prototype stages, but this is sometimes difficult. For example, despite all the submachine guns offered for sale or publicly displayed, a few completely unknown submachine guns appeared in the hands of troops in Chechnya. Who manufactured these weapons and their status remain mysteries to this day. It is difficult to determine from news photos, but the Chechnya submachine guns appear well made and to be production, rather than prototype, weapons. The story of submachine guns in Russia is thus evolving as of this writing.

A-9 AND A-7.62 SUBMACHINE GUNS

These two Tula-manufactured submachine guns are derived from the A-91 compact assault rifle and share a great deal of components with the assault rifle. The designations indicate the caliber of the two weapons: the A-9 is chambered in 9x19mm (9mm Parabellum) and the A-7.62 in 7.62x25mm. Oddly for a pistol-caliber firearm, these two submachine guns are gas operated. Gas operation adds expense to the manufacturing process, but this enables the weapon to fire from the closed bolt and reduces weight. Moreover, in the case of these two firearms, the production tooling for the A-91 is probably used to manufacture them, thus keeping costs to a minimum. According to Tula KBP, both have been adopted by the Russian MVD troops. This is very interesting, given the calibers for which the weapons are chambered. Chambering a new submachine gun in 9mm Parabellum indicates that the Russians may be transitioning into this caliber after more than 40 years of using the 9x18mm. It is possible that recent efforts to improve the perfor-

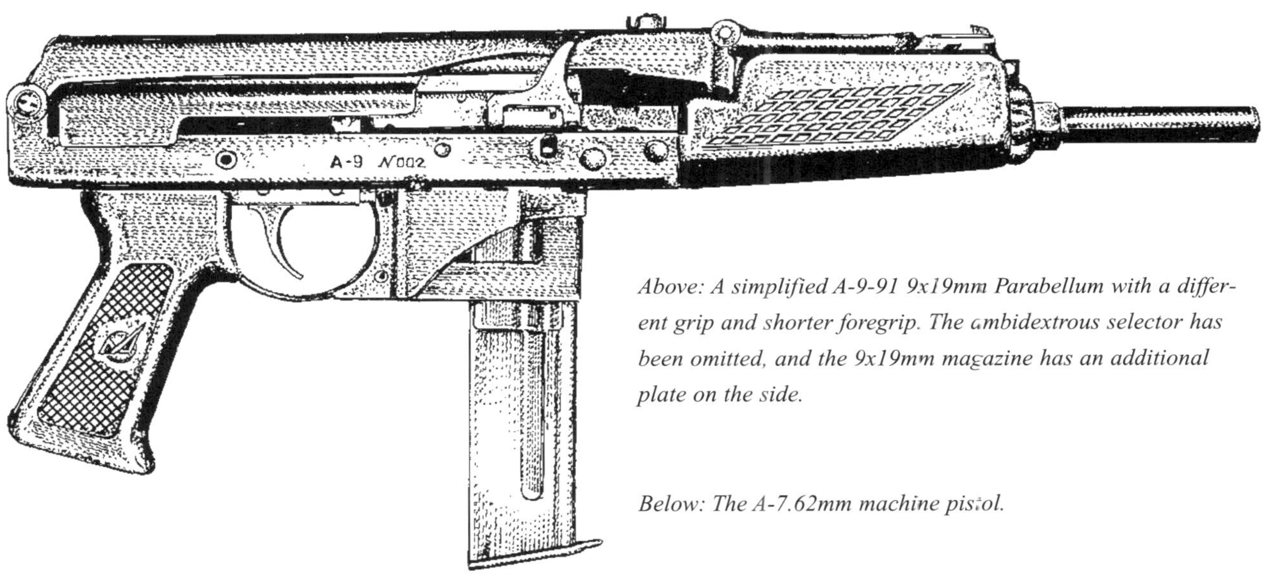

Above: A simplified A-9-91 9x19mm Parabellum with a different grip and shorter foregrip. The ambidextrous selector has been omitted, and the 9x19mm magazine has an additional plate on the side.

Below: The A-7.62mm machine pistol.

mance of the marginal 9x18mm may have been less than successful, hence the possible transition to 9x19mm in this and several other new Russian firearms for internal police and paramilitary use.

A-9 & A-7.62 SUBMACHINE GUN SPECIFICATIONS

Caliber	9x19mm/7.62x25mm
Operation	Gas, Select Fire
Weight	3.85 lbs. (1.75 kg.)
Length, Stock Extended	6.6 in. (167mm) (6.4 in./161mm)
Stock Folded	16.3 in. (415mm)
Effective Range	App. 100 m.
Magazine Capacity	20 or 30 Rounds
Sights	Front, Blade; Rear, Notch

AEK-919 AND AEK-919K *KASHTAN* (CHESTNUT)

The AEK-919 and its smaller AEK-919K are modern but thoroughly conventional submachine gun designs designed to an MVD requirement. Lower receivers are made of what appears to be carbon-fiber-reinforced polymer and upper receivers of stamped metal. The weapons are manufactured by the Kovrov Mechanical Plant Joint Stock Company, which is not a major Russian firearms manufacturer. The weapons have been shown alongside firearms of Tula KBP, so it is possible that there is a marketing arrangement between the two companies. It is likely that the company has the machine tool capability to manufacture simple firearms and thus was chosen to make some weapons under the old regime. Those examples of both weapons that have turned up in photographs have appeared to be well made, with component fit and finish on a par with Western weapons. Serial numbers have all been under 100, however, indicating that the firearms are prototype or preproduction weapons.

The design of both AEKs is clearly based on the Uzi, with a bolt that telescopes over the barrel. This design makes for a very compact weapon while at the same time allowing a relatively long barrel. The upper receiver of both AEKs is different; that of the AEK-919 is square in profile while that of the compact AEK-919K is rounded. The latter is probably to better enable the weapon to be concealed. As is that on the Uzi, the AEK-919/919K barrel is simply threaded into the

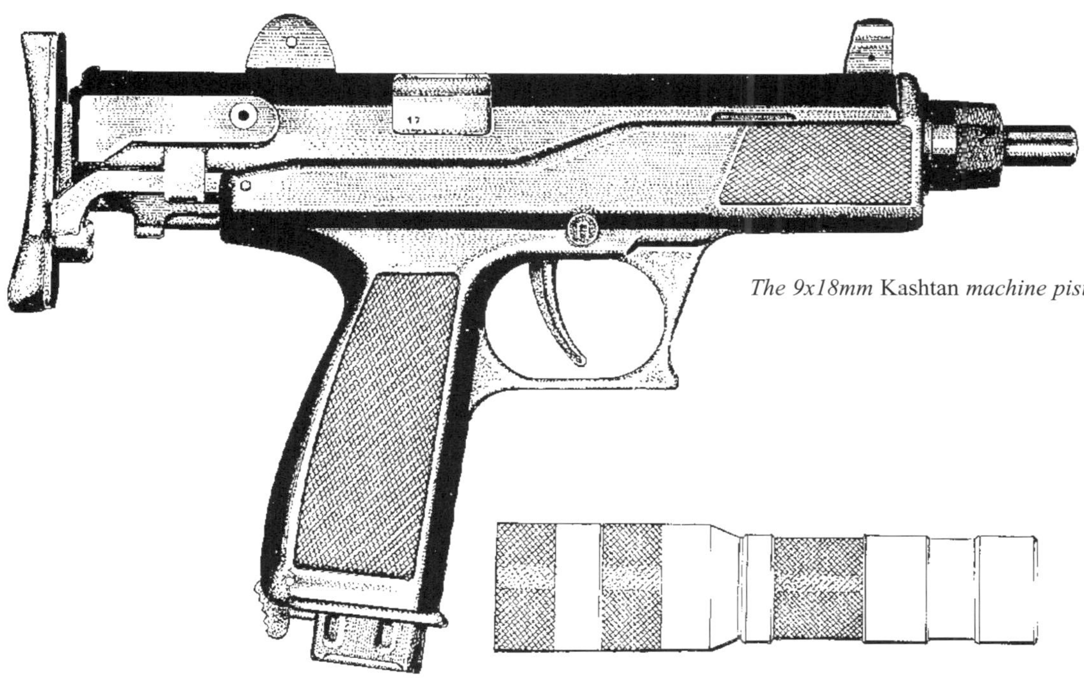

The 9x18mm Kashtan *machine pistol.*

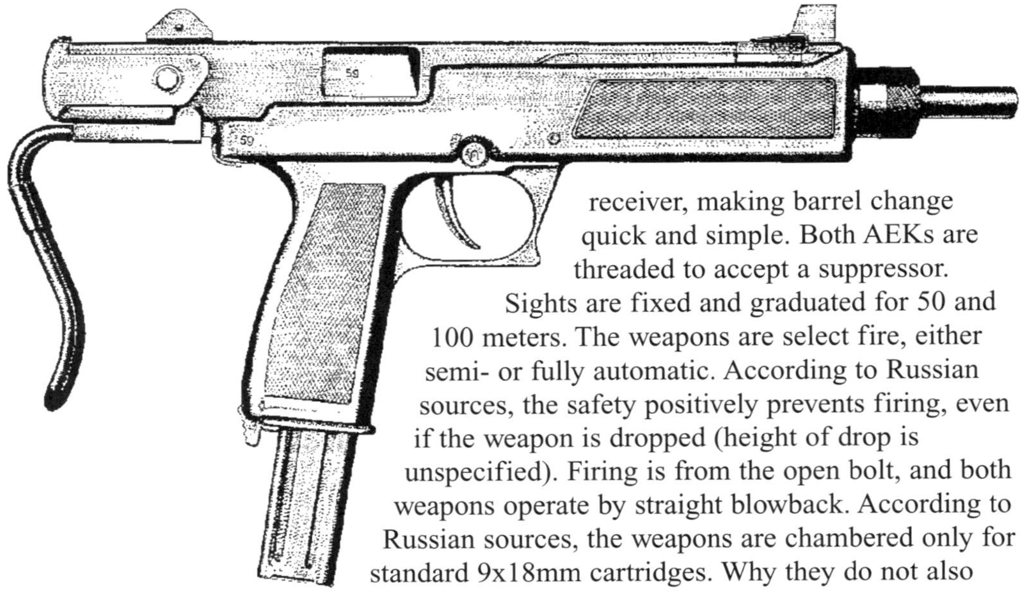

AEK-919 9x18mm submachine gun.

receiver, making barrel change quick and simple. Both AEKs are threaded to accept a suppressor.

Sights are fixed and graduated for 50 and 100 meters. The weapons are select fire, either semi- or fully automatic. According to Russian sources, the safety positively prevents firing, even if the weapon is dropped (height of drop is unspecified). Firing is from the open bolt, and both weapons operate by straight blowback. According to Russian sources, the weapons are chambered only for standard 9x18mm cartridges. Why they do not also fire the high-impulse 9x18mmM cartridges is a mystery, since the modification is a simple one.

AEK-919/AEK-919K SPECIFICATIONS

Caliber	9x18mm Makarov
Operation	Blowback, Select Fire
Weight	3.63 lbs. (1.65 kg.)
Length, Stock Extended	19.4 in. (495mm)
Stock Collapsed	12.7 in. (325mm)
Barrel Length	App. 7.5 in. (190.5mm)
Effective Range	App. 100 m.
Magazine Capacity	20 or 30 Rounds
Cyclic Rate	900 RPM
Sights	Front, Blade; Rear, Notch

BAKSAN (AKA "BUCKSUN")

Baksan literature has been shown only a few times, and there has never been any hardware displayed. The weapon exists in prototype only, and even this has never been publicly shown. The *Baksan* is almost certainly a candidate for a Russian personal defense weapon, a concept that has been proposed by several Western manufacturers over the past few years but that has yet to yield any real hardware other than the less than successful FN Herstal P90.

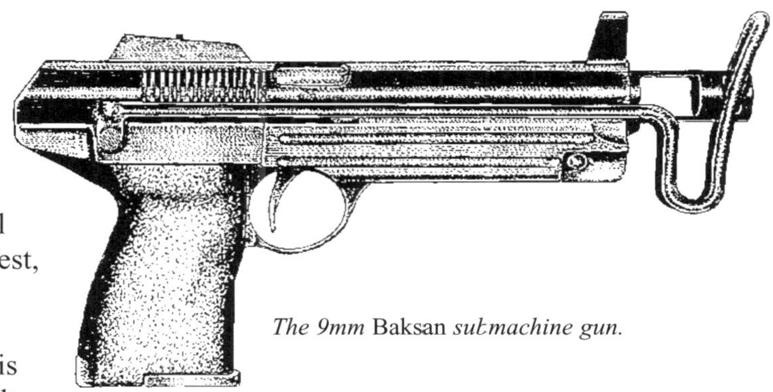

The 9mm Baksan *submachine gun.*

The designer, TsNIITochmash, has shown the weapon in computer drawings several times, along with tentative technical data to explore the market, but, as in the West, there appear to be no takers, although the *Baksan* does fire a standard cartridge. The problem with weapons such as the *Baksan* is that their configuration makes them difficult to control on full automatic, even in a low-impulse cartridge such as the 9x18mm PM. The *Baksan* is intended to fire the high-impulse 9x18mm PMM cartridge, the 9x19mm, and perhaps even the 9x21mm Russian, any of which would almost guarantee it to be extremely difficult to manage on full automatic, even with a folding stock and foregrip.

Nothing has been seen or heard of this weapon since 1995, so it is included here only for completeness.

BAKSAN SPECIFICATIONS (PROVISIONAL)

Caliber	9x18mmM; 9x19mm; 9x21mmRuss (?)
Operation	Delayed Blowback*
Weight	3.96 lbs. (1.8 kg.)
Length, Stock Extended	25.5 in. (650mm)
Stock Folded	13.7 in. (350mm)
Barrel Length	App. 8 in. (203mm)*
Effective Range	75–100 m.*
Magazine Capacity	20 Rounds*
Cyclic Rate	750–800 RPM
Sights	Front, Blade; Rear, Notch

*Assessed

BIZON (BISON), *BIZON*-2, AND *BIZON*-3

The *Bizon* was designed by a team headed by Viktor Kalashnikov and Alexey Dragunov, who continue in their fathers' footsteps. The younger Kalashnikov has reputedly essentially taken over the duties of his famed elder, who is now in his late 70s. Both the younger Kalashnikov and Dragunov work at Izmash Joint Stock Company. Like most other new Russian submachine guns, the *Bizon* was designed to an MVD requirement. The original *Bizon* was later designated *Bizon*-1 after the design was refined, improved, and developed into the *Bizon*-2 and the current *Bizon*-3.

The *Bizon* design is based on the AK-74M. The stock, receiver cover, and entire trigger mechanism directly interchange, and there is overall 60 percent parts interchangeability between the *Bizon* and the AK-74M. The major external difference is the 64-round helical magazine, which cannot be incorrectly loaded, according to Russian sources. Later versions of the *Bizon* have indicator holes in the magazine to show the user approximately how many rounds remain in the magazine. Openings are spaced at 4-, 24-, 44-, and 64-round increments.

Differences between the three versions of the *Bizon* are minor, but *Bizon*-3 is probably most different. The front sight on the *Bizon*-1 is Dragunov type, while that of the *Bizon-2* and *Bizon*-3 is Kalashnikov derived. Rear sights of the *Bizon*-1 and *Bizon*-2 are traditional tangent types with graduations at 50, 100, and 150 meters. *Bizon*-3 has an aperture rear sight. The handguard of the *Bizon* has continued to be refined and has differed slightly in each version. The *Bizon*-3 includes an integral mount for suppressors, compensators, and other muzzle devices. The

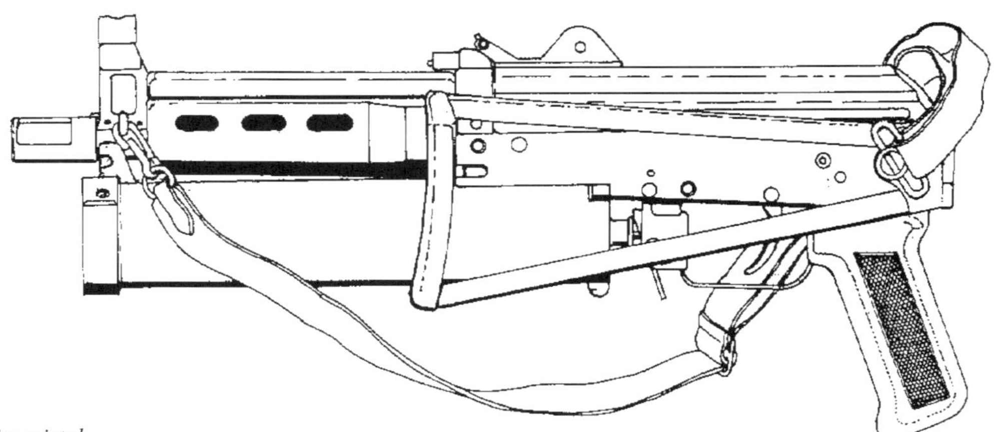

The 9mm Bizon-1 *machine pistol.*

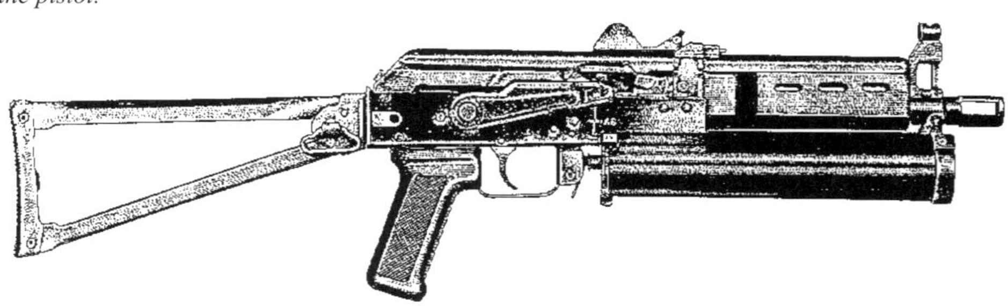

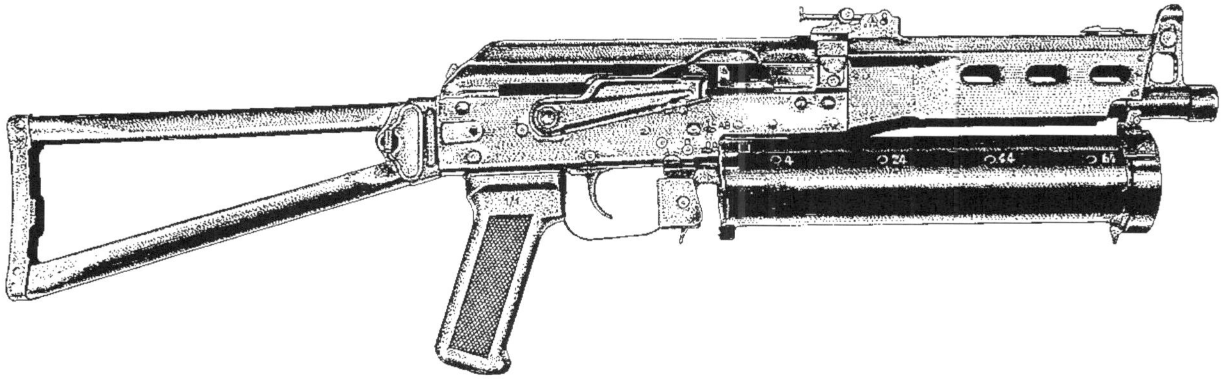

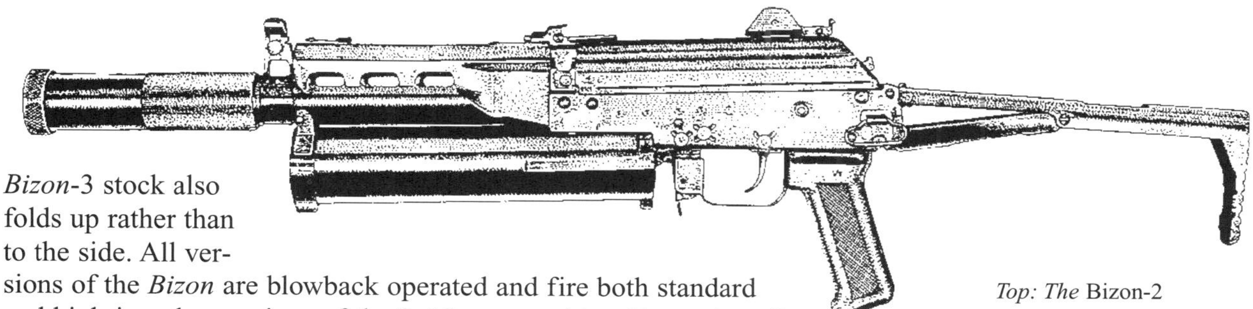

Bizon-3 stock also folds up rather than to the side. All versions of the *Bizon* are blowback operated and fire both standard and high-impulse versions of the 9x18mm cartridge. Examples of the *Bizon*-2 have been observed at arms shows with Western optical sights mounted.

Top: The Bizon-2 *9x18mm PMM submachine gun.*

Above: The Bizon-3 *9x18mm submachine gun with suppressor and fold-over stock.*

BIZON SPECIFICATIONS (*BIZON*-2 GIVEN, OTHERS SIMILAR)

Caliber	9x18mm; 9x18mmM
Operation	Blowback, Select Fire
Weight	5.5 lbs. (2.58 kg.)
Length, Stock Extended	25.9 in. (660mm)
Stock Folded	16.7 in. (425mm)
Barrel Length	9.4 in. (240mm)
Effective Range	App. 100 m.
Magazine Capacity	64 Rounds
Cyclic Rate	650–700 RPM
Sights	Front, Post; Rear, Notch

GEPARD (CHEETAH)

The experimental *Gepard* is yet another Kalashnikov-derived weapon, but a highly unusual one in that it is capable of firing no fewer than five different cartridges with minor modification. It is manufactured by a small specialist Moscow company, Rex Firearms, which possibly is affiliated with the Russian firearms giant, Izmash. This particular weapon was derived from the AKS-74U, with which it shares 65 percent parts commonality. The *Gepard* fires 9x17mm (.380ACP), 9x18mm (both varieties), 9x19mm, 9x21mm Russian, and a new 9x30mm Russian cartridge. Presumably the *Gepard* is also capable of firing Western 9x21mm. If so, this makes the weapon capable of firing six different cartridges!

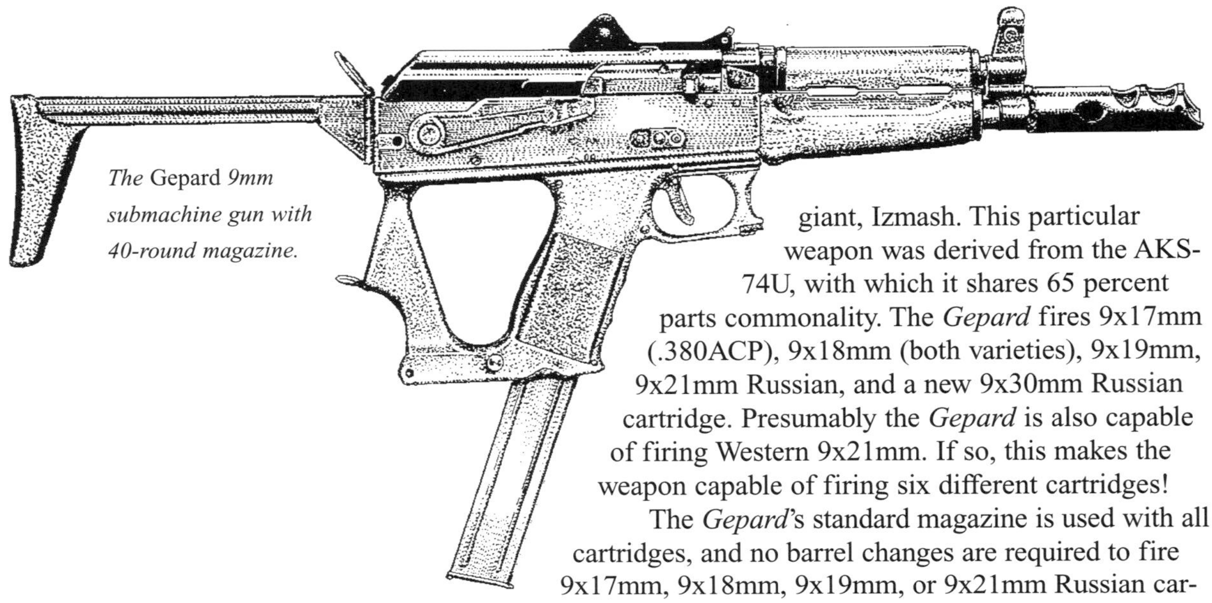

The Gepard *9mm submachine gun with 40-round magazine.*

The *Gepard*'s standard magazine is used with all cartridges, and no barrel changes are required to fire 9x17mm, 9x18mm, 9x19mm, or 9x21mm Russian cartridges. To fire the 9x30mm cartridge, the weapon's chamber is changed via a threaded sleeve that screws into the barrel extension. To fire the other cartridges, however, matters get a bit complicated. The operating system of the *Gepard* may be blowback, blowback with internal buffers, gas-assisted blowback with internal buffers, delayed blowback in any of the foregoing configurations, or gas-operated locked breech, depending upon which cartridge is being fired. Russian articles on the *Gepard* state that blowback operation is used with 9x17mm, 9x18mm, and 9x19mm cartridges. Delayed blowback, gas-assisted blowback, or blowback with buffers is used with 9x21mm. Gas operation with a locked breech is reserved for the 9x30mm, whose bullet leaves the muzzle at 1,970 fps. Change in operation is achieved by switching modular bolt groups and recoil spring groups, which though seemingly simple is actually very complicated for a small arm. The muzzle is threaded and may be fitted with a variety of devices, such as brakes and suppressors.

Despite the fact that this weapon is a technical achievement of the first order, it is a weapon that probably will never see full production; the complexity involved in changing it from one caliber to another overcomes whatever flexibility multicaliber capability achieves. The *Gepard* has no fewer than three different bolt groups and two recoil spring mechanisms with which the operator must contend. Moreover, the Russians themselves have weapons in production that fulfill the missions for which this unusual submachine gun is designed. Despite the fact that

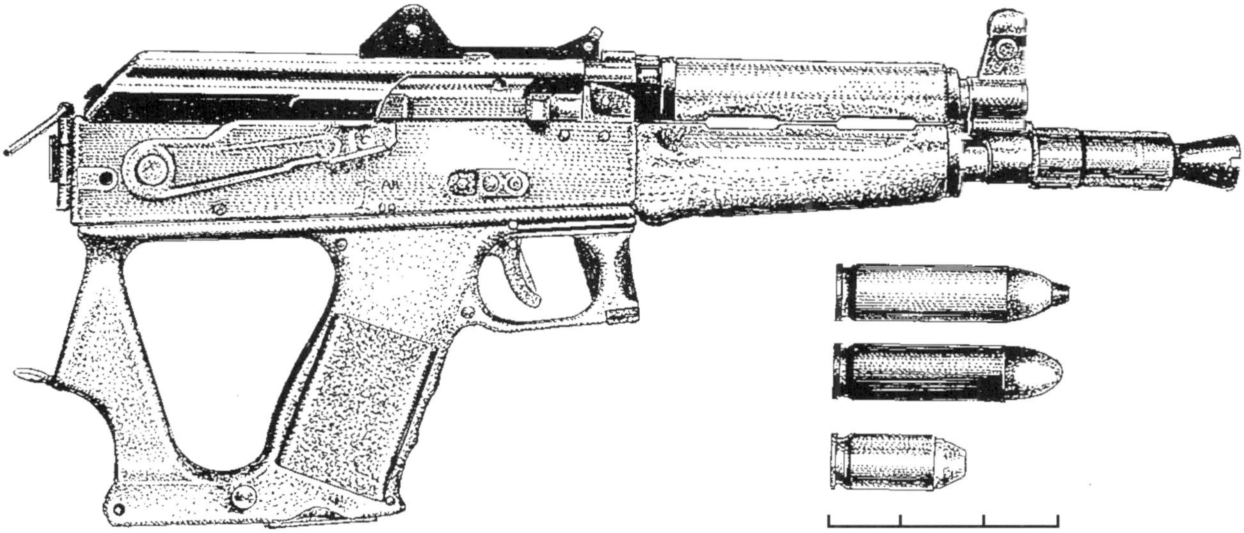

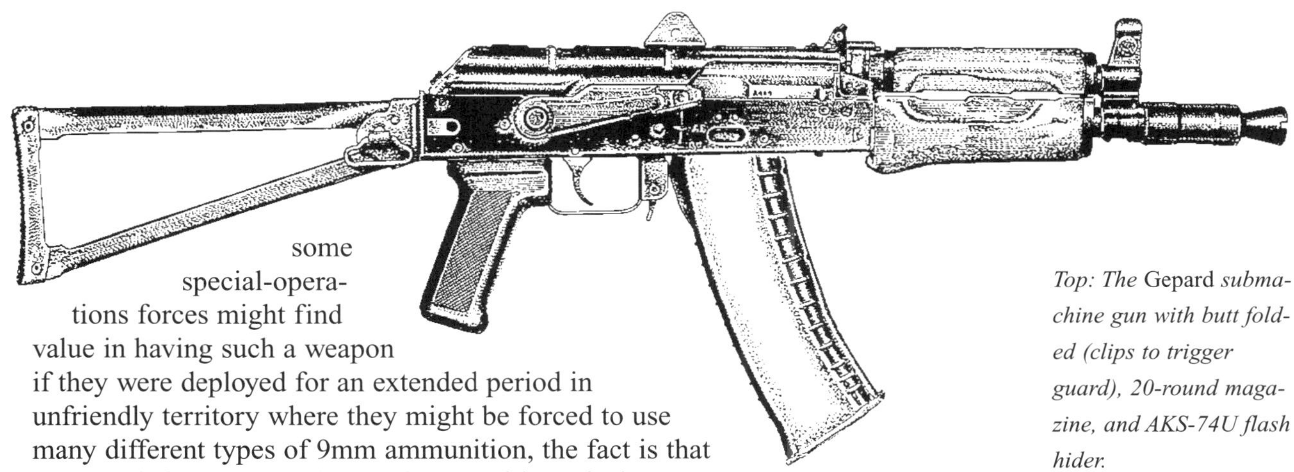

Top: The Gepard *submachine gun with butt folded (clips to trigger guard), 20-round magazine, and AKS-74U flash hider.*

Above: The latest AKS-74U shortened assault rifle.

some special-operations forces might find value in having such a weapon if they were deployed for an extended period in unfriendly territory where they might be forced to use many different types of 9mm ammunition, the fact is that most such forces have plenty of ammunition of whatever type they want. Moreover, 9x19mm ammunition is dirt-common virtually everywhere except Russia and former Warsaw Pact countries, where 9x18mm is the norm. In point of fact, like all such firearms, the operational disadvantages of the *Gepard* far offset any other advantages that its multicaliber capability may afford. In addition, the user must carry and keep track of those bolt and recoil spring groups. The *Gepard* is simply too complicated to be used by military forces in its current configuration.

GEPARD SPECIFICATIONS

Caliber	9x18mm; 9x18mmM; 9x19mm; 9x21mm; 9x30mm
Operation	Blowback/Delayed Blowback/Gas, Select Fire
Weight	4.2 lbs. (2.0 kg.)
Length, Stock Extended	25.1 in. (640mm)
Stock Folded	16.5 in. (420mm)
Barrel Length	9.2 in. (235mm)
Effective Range	100–200 m.
Magazine Capacity	22 or 40 Rounds
Cyclic Rate	650–700 RPM, Depending on Caliber
Sights	Front, Post; Rear, Notch

PP-91 *KEDR* (CEDAR) AND *KLIN* (WEDGE)

As mentioned above, the *Kedr/Klin* designs date to the 1960s, but the *Kedr* was not placed into production at that time. By the late 1980s, however, the *Kedr*'s time had come, and not only was it modernized and placed into series production in 1993, but an upgraded version called the *Klin* was developed and produced beginning in mid-1994. Both weapons are currently in use by the MVD. The only difference between the two submachine guns is that the *Klin* is capable of firing both standard and high-impulse versions of the 9x18mm cartridge. As a result, the Klin has a slightly higher cyclic rate when firing the 9x18mm PMM cartridge.

The *Klin* is no more than an improved version of the *Kedr*. Development of the *Klin* was undertaken because the Russians found that the 9x18mm PM round was ineffective at ranges greater than 50 meters. Accordingly, they developed a "modernized" version of the 9x18mm with 9x19mm performance. This cartridge is designated the PMM and can be differentiated from its predecessor by the conical ogive of the bullet. Presumably, the 9x18mm PMM will chamber in standard 9x18mm firearms, because case length of the two rounds is identical. Russian literature makes it clear that weapons chambered for the new high-impulse 9x18mm can chamber and fire standard cartridges, but older weapons cannot fire the new cartridge because of its higher chamber pressures. Regardless of the operational problems that the 9x18mm PMM cartridge will almost surely cause, it gives the Russians a cartridge that is squarely in the 9x19mm performance envelope. Why they did not simply design the *Klin* to fire both 9x19mm and 9x18mm cartridges remains

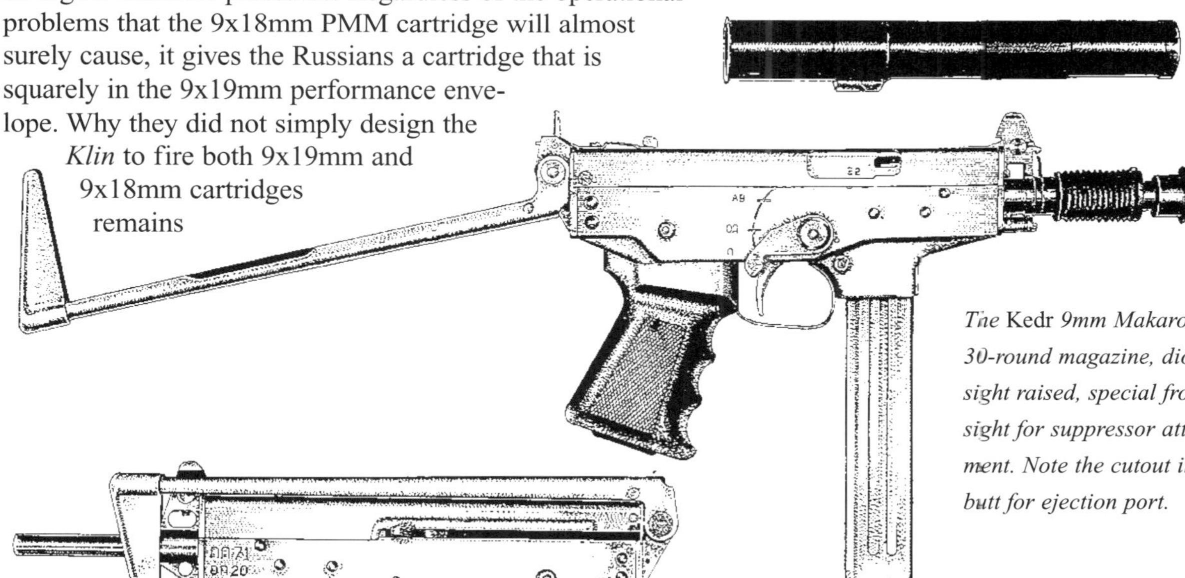

The Kedr *9mm Makarov with 30-round magazine, diopter sight raised, special front sight for suppressor attachment. Note the cutout in the butt for ejection port.*

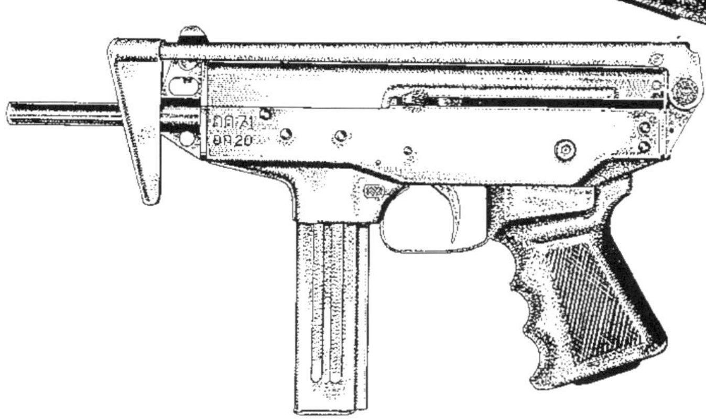

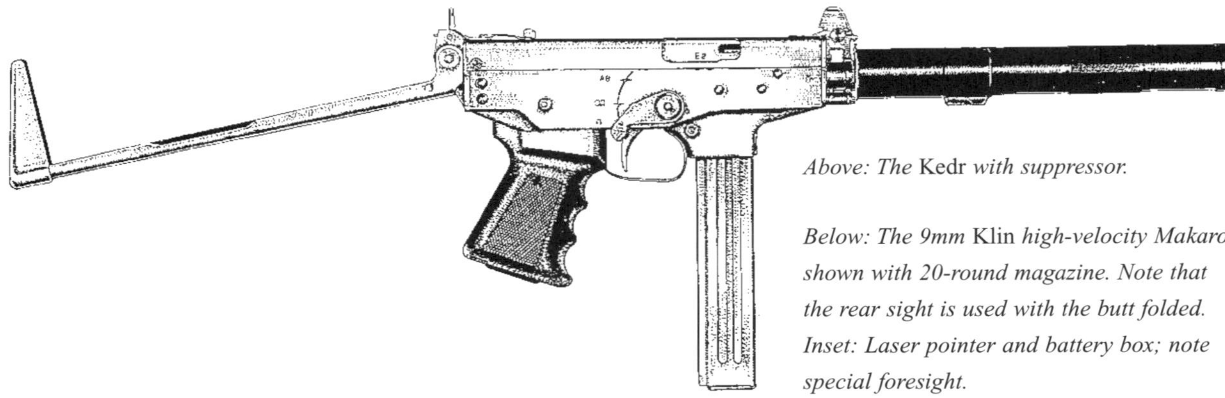

Above: The Kedr *with suppressor.*

Below: The 9mm Klin *high-velocity Makarov shown with 20-round magazine. Note that the rear sight is used with the butt folded. Inset: Laser pointer and battery box; note special foresight.*

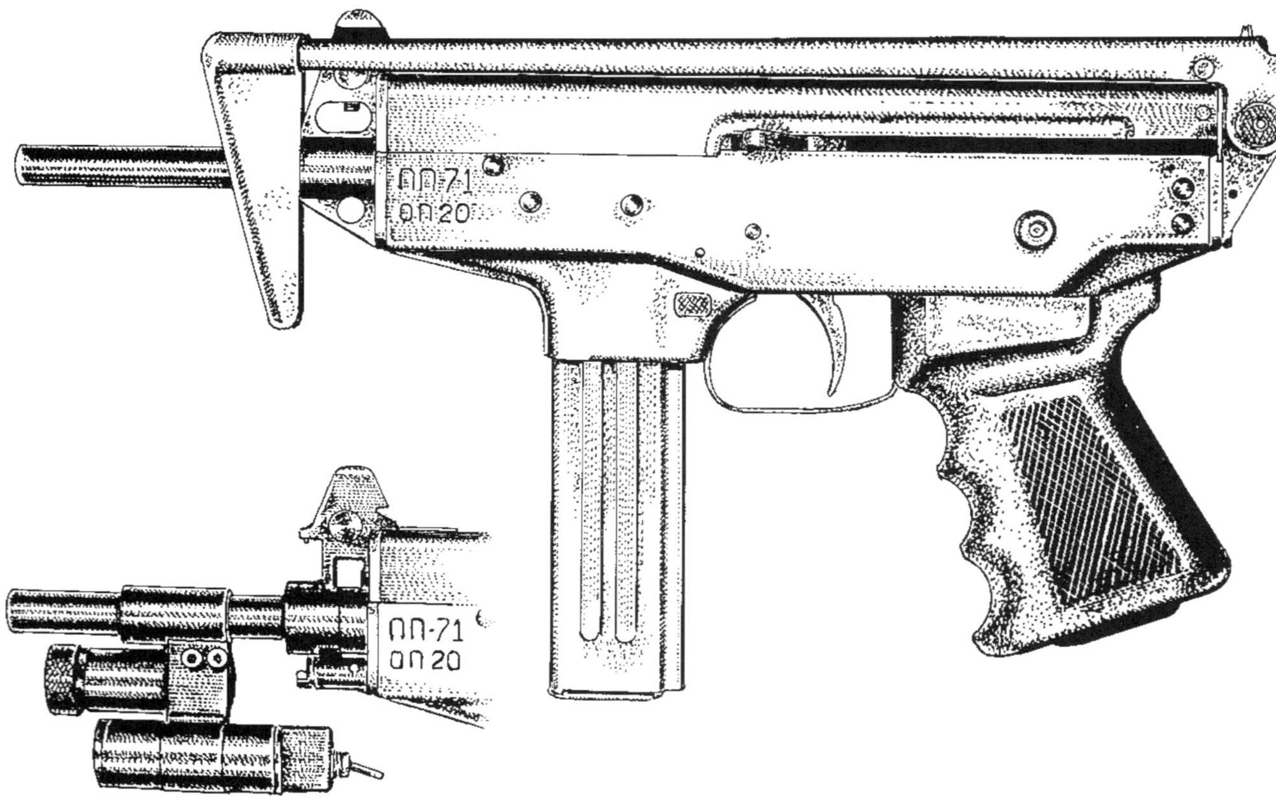

a mystery. They have done this in other submachine guns, as we shall presently see. The *Klin* will probably eventually replace the *Kedr*: the two weapons are identical in size and operation, and the *Klin* has far better effectiveness than the earlier weapon, although as of early 1998, both remain in production.

Both weapons are blowback operated. *Kedr* is simple blowback, whereas *Klin* is delayed blowback. This is accomplished by the simple expedient of machining a spiral groove into the chamber to hold the cartridge case in place long enough for the chamber pressure to drop to a safe level before the casing is extracted and ejected. The system is virtually identical to that of the PMM Makarov pistol. Both *Kedr* and *Klin* are select fire (semi- and full automatic), fire from the closed bolt, and employ a hammer fire system. A variety of accessories is available for these little submachine guns, including suppressors and laser sights (the latter are produced in Belarus). The only way to externally distinguish the Kedr and Klin is through the markings on the right side of the receiver, although there are a few minor internal differences.

KEDR AND *KLIN* SPECIFICATIONS

	Kedr	***Klin***
Caliber	9x18mm	9x18mm and 9x18mmM
Operation	Blowback	Delayed Blowback
Weight	4 lbs. (1.82 kg.)	4.02 lbs. (1.83 kg.)
Length, Stock Extended	21.25 in. (540mm)	21.25 in. (540mm)
Stock Folded	11.8 in. (300mm)	11.8 in. (300mm)
Barrel Length	4.7 in. (120mm)	4.7 in. (120mm)
Effective Range	App. 100 m.	App. 100 m.
Cyclic Rate	800–850 RPM	1,080–1,200 RPM
Sights	Front, Blade; Rear, Notch	Front, Blade; Rear, Notch

OTs-02/PP-891 *KIPARIS* (CYPRESS)

The Tula KBP-manufactured *Kiparis* is described in Russian literature as a weapon for use by MVD troops fighting organized crime. The weapon is in use by many MVD units and was observed in use by Russian units during the Chechnya fighting. It appears to be a well-made submachine gun, and it is apparently popular with the troops who use it. The *Kiparis* also appears in the *Russian Military Catalog*, indicating that it is available for export.

Other than the fact that it appears well finished and functional, the *Kiparis* is a conventional design, similar both in concept and execution to the *Kedr/Klin* described previously. Like the *Kedr* and *Klin*, the *Kiparis* was a competitor in the 1960s submachine gun competition. The design was actually finalized in 1972, but the gun was never produced. However, rampant criminal activities of the 1990s revived interest in submachine guns of all types, and the *Kiparis* design was resurrected and placed into production.

The *Kiparis* is blowback operated and has select fire. Per the Tula specifications, it fires from the closed bolt, thus enhancing accuracy. Indeed, Tula KBP sales literature claims that the little submachine gun is extremely accurate in semi-automatic fire. The trigger mechanism was modeled after that of the Tokarev TT-33 pistol and is removed from the *Kiparis* en bloc very much like that of the TT-33. It has a collapsible stock, which folds over the top of the weapon. The *Kiparis* is chambered for the standard 9x18mm cartridge and is equipped with a suppressor and laser-aiming module as standard equipment. According to Russian literature, however, the *Kiparis* can easily be modified to use the more powerful

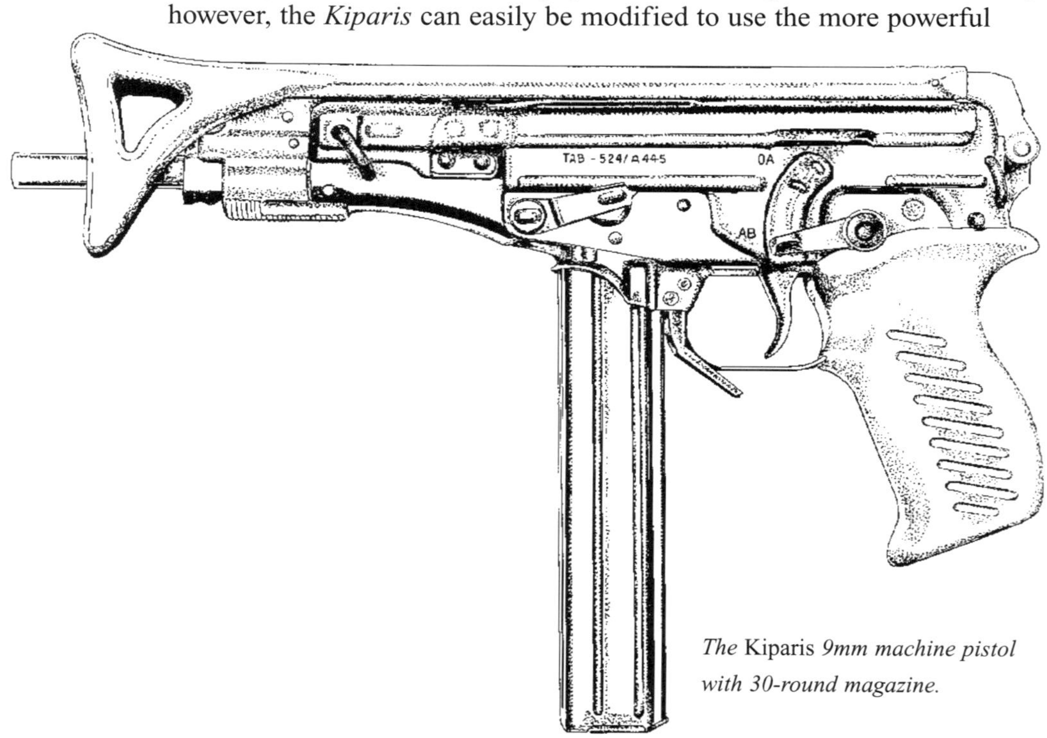

The Kiparis *9mm machine pistol with 30-round magazine.*

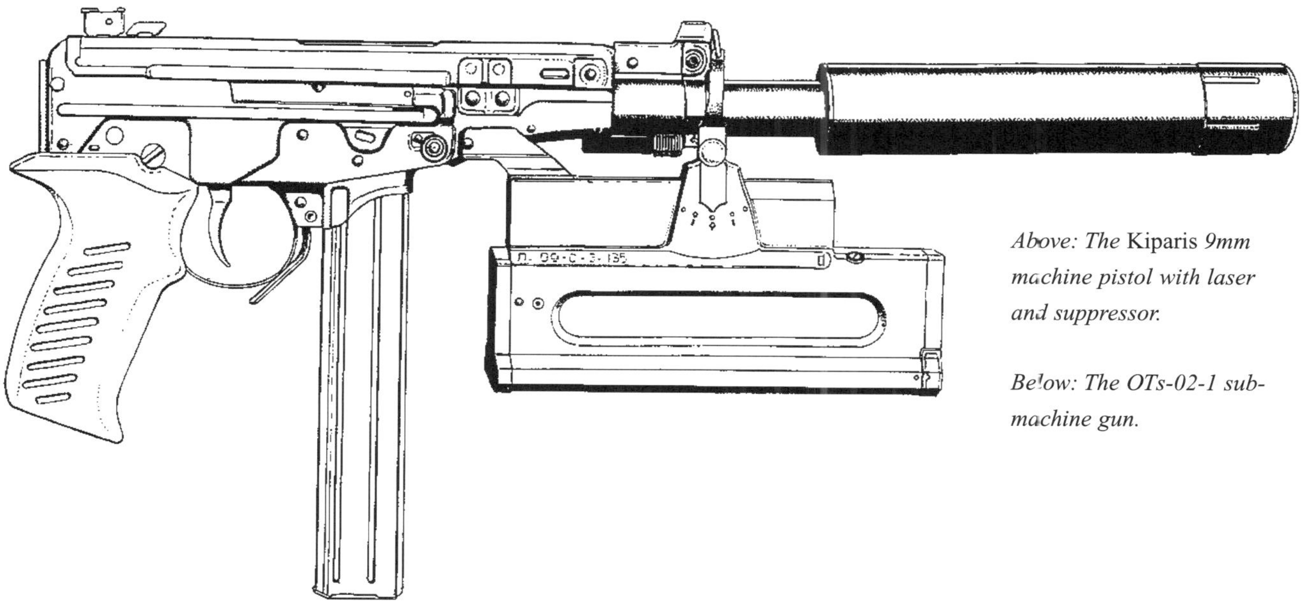

Above: The Kiparis *9mm machine pistol with laser and suppressor.*

Below: The OTs-02-1 submachine gun.

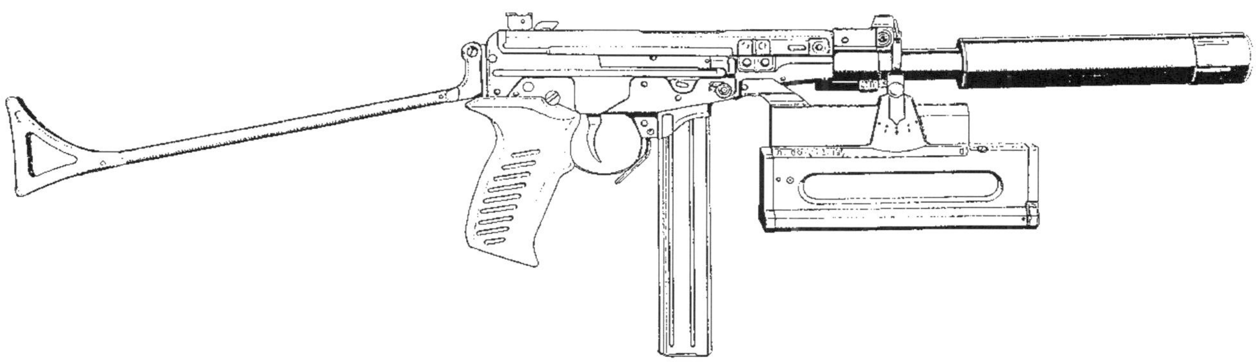

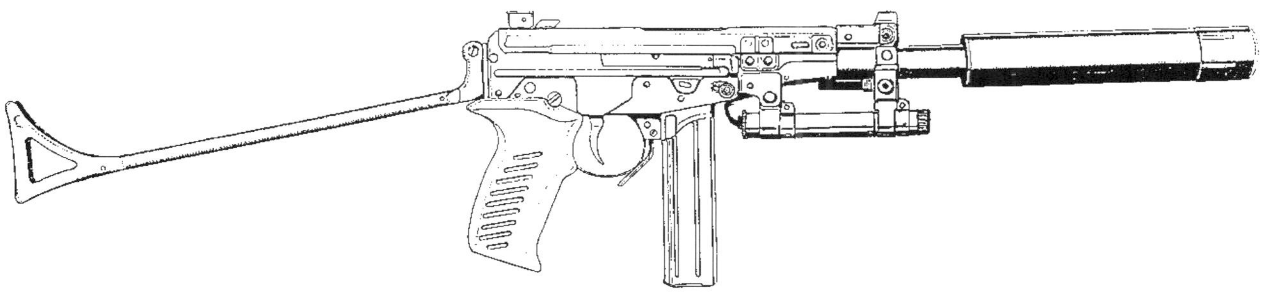

PMM 9x18mm cartridge. There are three magazines available for the *Kiparis*: 10, 20, and 30 rounds. A 10-round magazine may seem small for a fully automatic submachine gun, but it was a specific requirement by the MVD. Service life of the *Kiparis*, as for many other recent Russian small arms, is stated to be low: only 6,000 rounds. It will be noted that several of the weapons we have discussed here have had claimed service lives that are far below what would be expected in similar Western weapons. Whether this is conservatism by the Russians or simply that they intentionally manufacture some weapons to wear out in fairly short order is unknown. The Russians are certainly capable of manufacturing firearms that are the equal of any in the world in terms of service life, reliability, and ruggedness.

OTs-02 *KIPARIS* SPECIFICATIONS

Caliber	9x18mm
Operation	Blowback, Select Fire, Closed Bolt
Length, Stock Extended	23.6 in. (600mm)
Stock Folded	12.4 in. (317mm)
Barrel Length	App. 10 in. (394mm)
Effective Range	50–100 m.
Magazine Capacity	10, 20, or 30 Rounds
Cyclic Rate	800 RPM
Sights	Front, Protected Blade; Rear, Protected Notch, Adjustable

OTs-22 SUBMACHINE GUN

The OTs-22 is an extremely compact and lightweight submachine gun design from Tula KBP in 9x19mm caliber that is possibly derived in part from the Ingram submachine gun. It also bears a very close resemblance to the AEK-919K submachine gun described earlier. There is little else known regarding this submachine gun. The OTs-22 is blowback operated, fires from the open bolt, and has ambidextrous controls and a grip safety. There are two magazines available in either 20- or 30-round capacity. Presumably, the 20-round magazine is for concealed carry, as the little gun is certainly small enough to be carried clandestinely.

OTs-22 SPECIFICATIONS

Caliber	9x19mm
Weight	2.86 lbs. (1.3 kg.)
Length, Stock Folded	9.8 in. (250mm)
Stock Extended	18 in. (460mm)
Cyclic Rate	800–900 RPM
Effective Range	50 m.
Magazine Capacity	20 or 30 Rounds

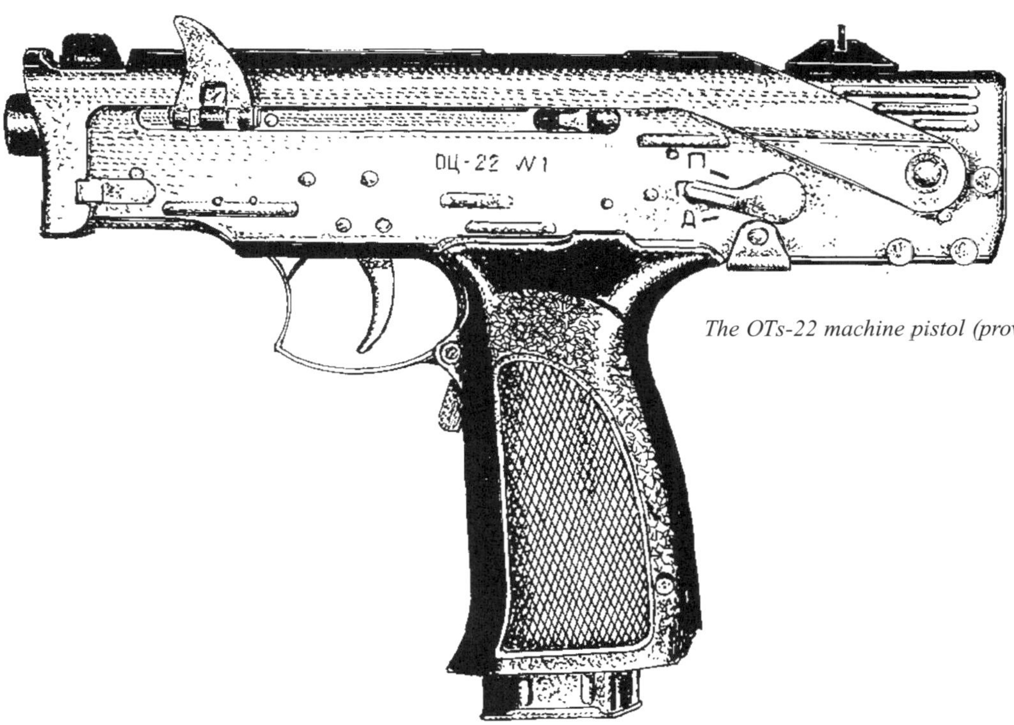

The OTs-22 machine pistol (provisional drawing).

PP-90/PP-90M/PP-90M1 9x18mm/9x19mm SUBMACHINE GUN

The PP-90 is a very compact collapsible submachine gun. It is designed for MVD troops engaged in special operations. Given its extremely compact size and ability to be brought into action (according to the manufacturer) in as little as two seconds after training, the PP-90 is also ideal for executive protection situations where a low profile is mandatory. When folded, the PP-90 is compact and lightweight, and, most important for maintaining a low profile, it does not look like a gun, if the operative happens to expose it by accident.

The PP-90 is blowback operated and uses advanced primer ignition and a buffer mechanism to optimize accuracy and reduce muzzle climb. The firing pin strikes the primer of the cartridge while 1.3mm of the cartridge case is still outside the chamber. This delays opening the breech by making use of the forward inertia of the bolt to help seal the breech and also enables the mass of the bolt to be reduced somewhat. The mass of the bolt actually is above the barrel, which enables the recoil forces to be directed below the point where the PP-90 is placed on the shoulder. As mentioned previously, this design feature minimizes muzzle climb on fully automatic weapons. There is no provision for select fire on the PP-90; it is fully automatic only, although after training the cyclic rate is such that operators should be able to squeeze off short bursts without difficulty. According to the manufacturer, it is physically impossible for the PP-90 to be fired in a partly unfolded configuration. The weapon must be fully unfolded and locked in position to fire. There is also an automatic safety mechanism that positively prevents firing if

The 9mm special small submachine gun PP-90. Top: Stock extended and sights raised. Left: Top view. Right: Magazine recoil spring and guide bolt; weapon folded, safety catch on "fire."

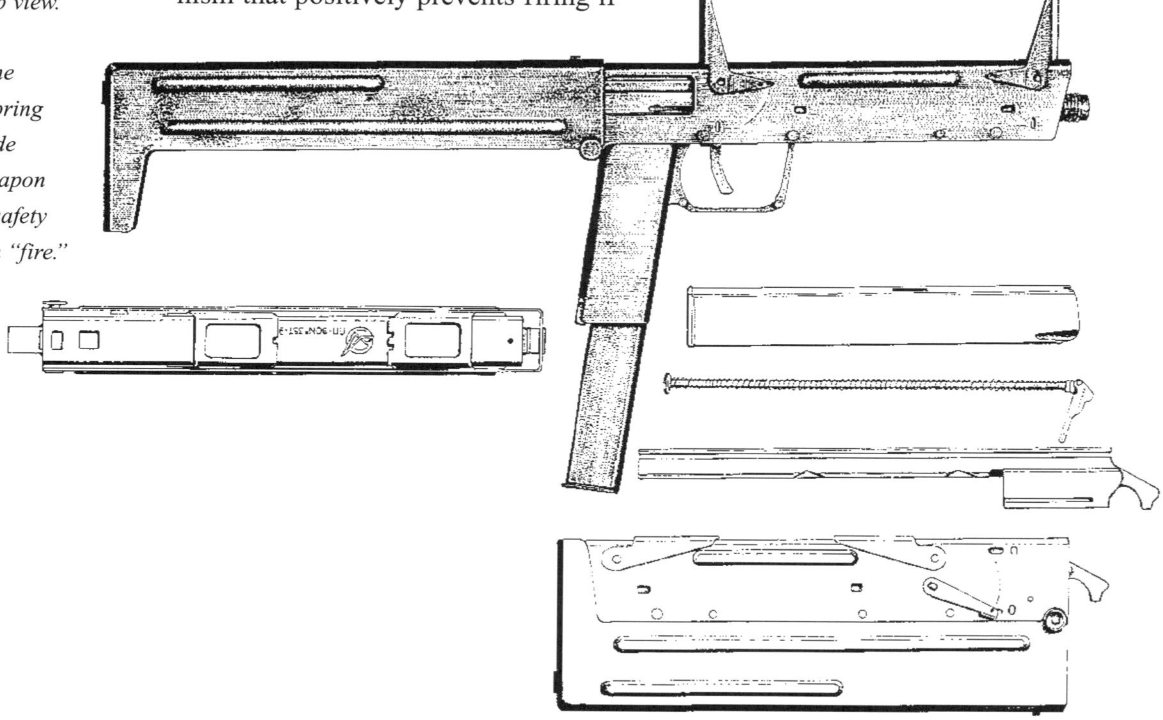

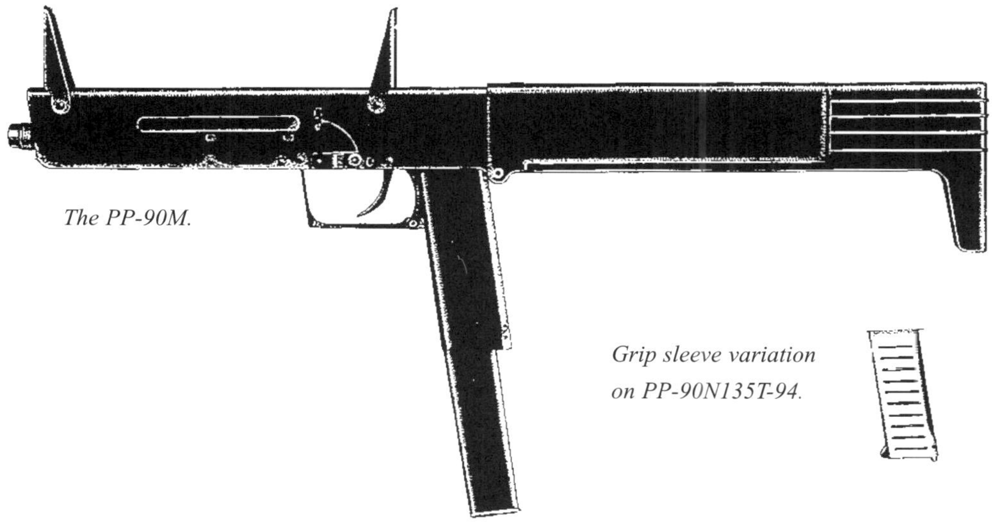

The PP-90M.

Grip sleeve variation on PP-90N135T-94.

PP-90M1 9x19mm Parabellum folded.

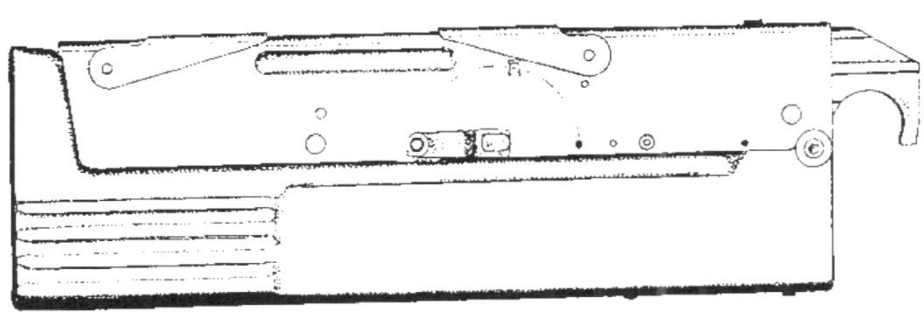

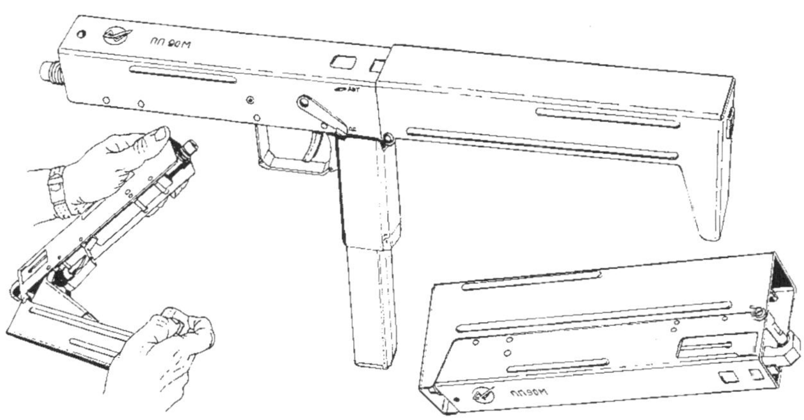

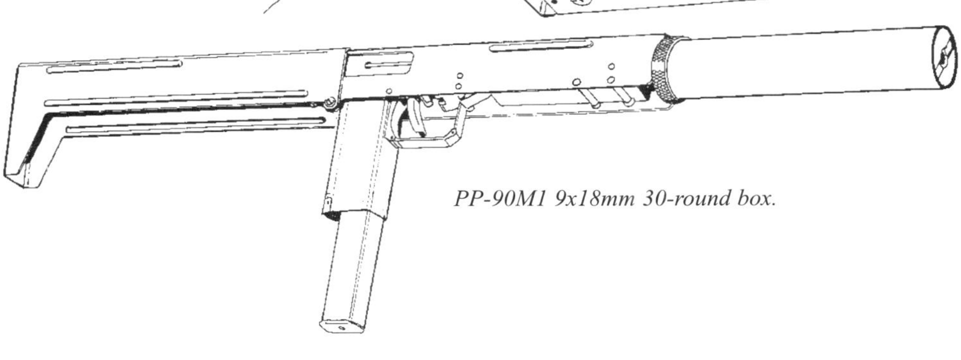

PP-90M1 9x18mm 30-round box.

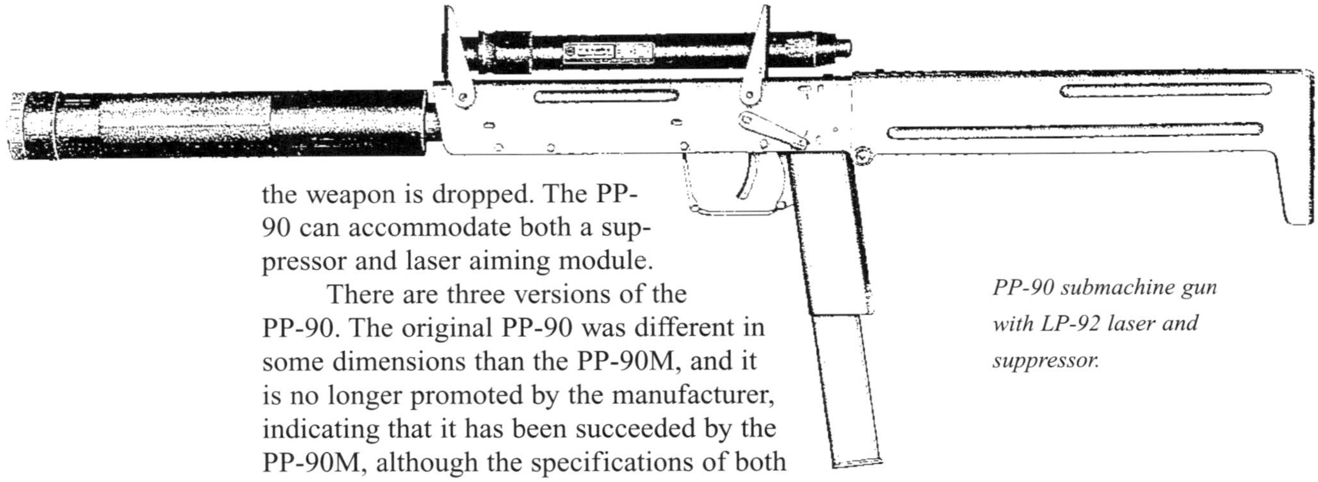

PP-90 submachine gun with LP-92 laser and suppressor.

the weapon is dropped. The PP-90 can accommodate both a suppressor and laser aiming module.

There are three versions of the PP-90. The original PP-90 was different in some dimensions than the PP-90M, and it is no longer promoted by the manufacturer, indicating that it has been succeeded by the PP-90M, although the specifications of both will be listed below. The PP-90M1 is virtually identical to the PP-90M, save for the fact that it is chambered in 9x19mm.

PP-90/PP-90M/PP-90M1 SPECIFICATIONS

	PP-90M (PP-90)	**PP-90M1**
Caliber	9x18mm	9x19mm
Operation	Blowback, Full Auto	Blowback, Full Auto
Weight	2.6 lbs. (1.42 kg.) (3.9 lbs./1.83 kg.)	3.2 lbs. (1.45 kg.)
Length, Stock Folded	11 in. (280mm) (10.6 in./270mm)	10.8 in. (276mm)
Stock Extended	19 in. (485mm) (19.2 in./490mm)	19 in. (485mm)
Barrel Length	App. 8 in. (203mm)	App. 8 in. (203mm)
Effective Range	App. 100 m.	App. 100 m.
Cyclic Rate	600–800 RPM	600–800 RPM
Magazine Capacity	30 Rounds	30 Rounds
Sights	Front, Blade; Rear, Notch	Front, Post; Rear, Notch

PP-93 9x18mm SUBMACHINE GUN

The PP-93 can be considered essentially a nonfolding version of the PP-90 described above. Both submachine guns are manufactured by Tula KBP, and both are virtually identical in function. For example, both use advanced primer ignition, both have a recoil buffer, both have an automatic safety to prevent firing if the weapon is dropped, and both can be fitted with essentially identical laser aiming devices and suppressors.

There are three primary differences between the PP-93 and its folding cousin. The first is that the stock of the PP-93 does not offer as favorable a position to offset recoil forces as the PP-90. The stock of the PP-93 folds forward across the top of the weapon, and when it is extended it locks into place so that it meets the shooter's shoulder below the line of recoil forces, thereby enhancing the tendency of the weapon to climb when fired on full automatic. The second difference is that the PP-93 is select fire, rather than full auto only. The third difference is that the PP-93 apparently is chambered only in both versions of the 9x18mm. Why Tula chose to chamber the PP-90 in 9x19mm and not do the same with the virtually identical PP-93 is another of those questions that for the moment remains unanswered.

The PP-93 is approximately the size of a very large pistol with the stock folded, and it is therefore somewhat concealable for clandestine operations. In fact, Tula provides a shoulder holster for the PP-93 as a part of the accessory kit that accompanies it, although there is no mistaking the PP-93 for what it is! Like the rest of the Tula-manufactured firearms, the PP-93 is claimed to be in service with MVD forces, which clearly must have a tremendous requirement for submachine guns, given the number and variety of them that are currently being manufactured in the Russian Federation.

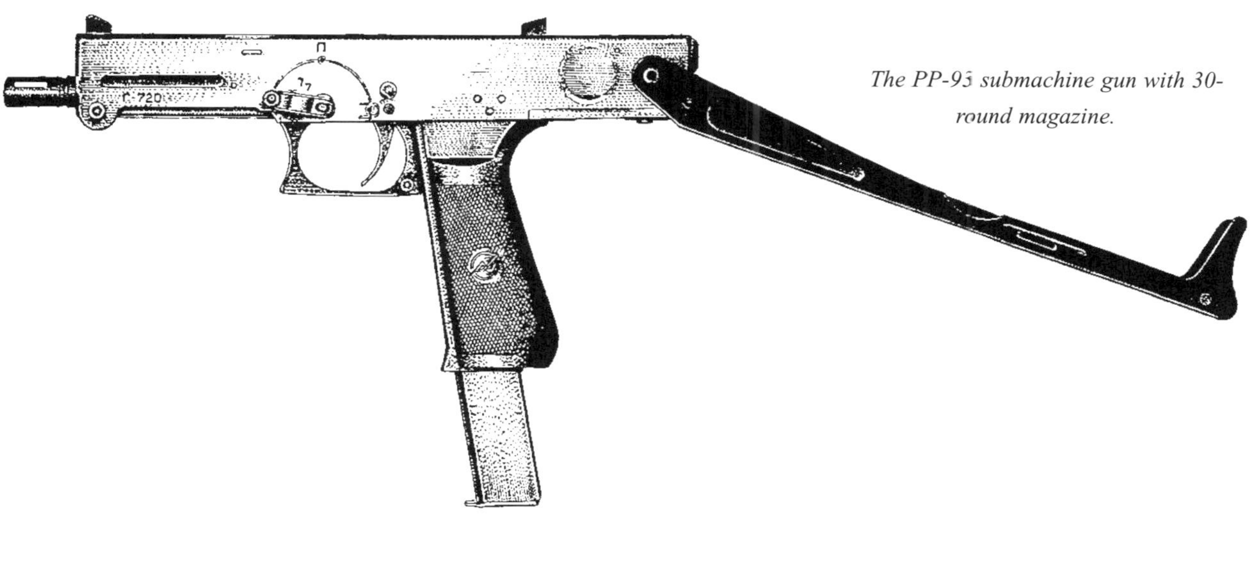

The PP-93 submachine gun with 30-round magazine.

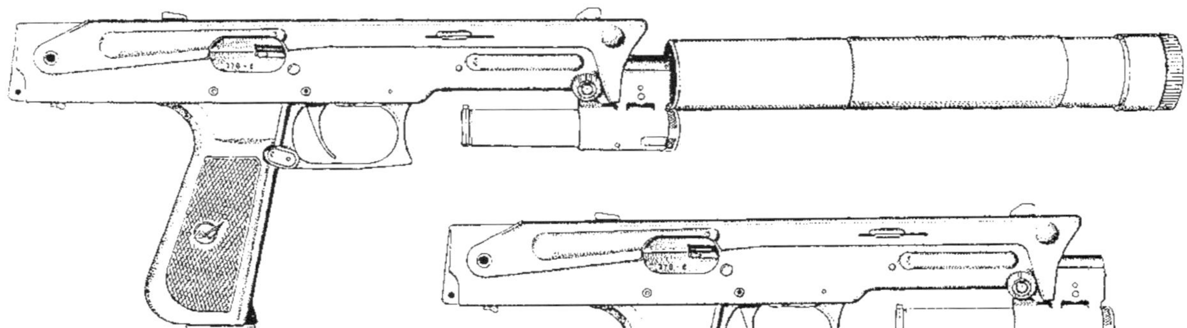

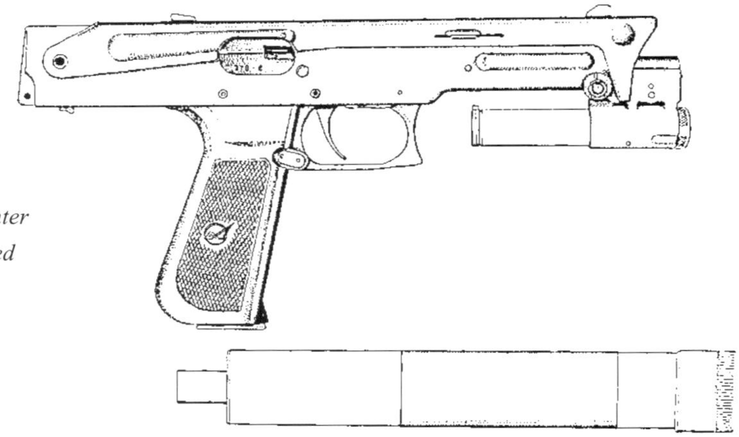

The PP-93 submachine gun with laser pointer and suppressor. (The suppressor is also used with PP-90M.)

PP-93 SPECIFICATIONS

Caliber	9x18mm/9x18mmM
Operation	Blowback, Select Fire
Length, Folded	12.7 in. (325mm)
Extended	21.5 in. (547mm)
Barrel Length	App. 8 in.
Effective Range	50–75 m.
Cyclic Rate	700 RPM
Magazine Capacity	20 or 30 Rounds
Sights	Front, Protected Blade; Rear, Adjustable Notch

CHAPTER FOUR

TACTICAL SHOTGUNS

Whether it is called a combat shotgun, a fighting shotgun, or a tactical shotgun, this firearm has become an accepted battlefield and law enforcement weapon throughout the world. The shotgun as a battle weapon was pioneered in the United States, where it served with distinction in every war fought by Americans, beginning with the Revolution. It was not until World War I that the United States' adversaries discovered, much to their dismay, what a devastating and versatile battlefield weapon the shotgun can be. The shotgun was used by U.S. Army "doughboys" with such great effect against the Germans in the trenches of France that the German government registered a formal complaint against the United States, claiming that the Winchester M1897s that formed the bulk of the U.S. tactical shotguns of the day were "inhumane." The claim got nowhere, and shotguns have been on the battlefield ever since, mostly in the hands of Americans.

In recent years, however, other nations have begun to produce tactical shotguns, probably driven by developments in shotgun ammunition that make these firearms even more flexible and adaptable to battlefield conditions. Another factor driving shotgun development has been the plethora of military operations in urban terrain (MOUT) battles in recent years and growing prospects for even more MOUT operations in the future. It is in such close terrain as cities and jungles that the shotgun comes into its own.

In peacekeeping operations, it can be loaded with nonlethal ammunition and quickly changed over to lethal ammunition if necessary. Current nonlethal shotgun loads include gas rounds, "bean bags," rubber "stingballs," rubber buckshot, rubber batons, "flash-bangs," and more. Lethal rounds include barrier penetration rounds, antivehicular rounds, "door breakers," flechettes, and much more. In development are small high-explosive rounds and mini shaped charges to defeat light armor.

The Chechnya unpleasantness taught the Russian military a number of lessons besides how unwise it is to employ tanks without infantry support, especially when the enemy can get up on rooftops and shoot down onto the relatively unprotected tops of tanks.

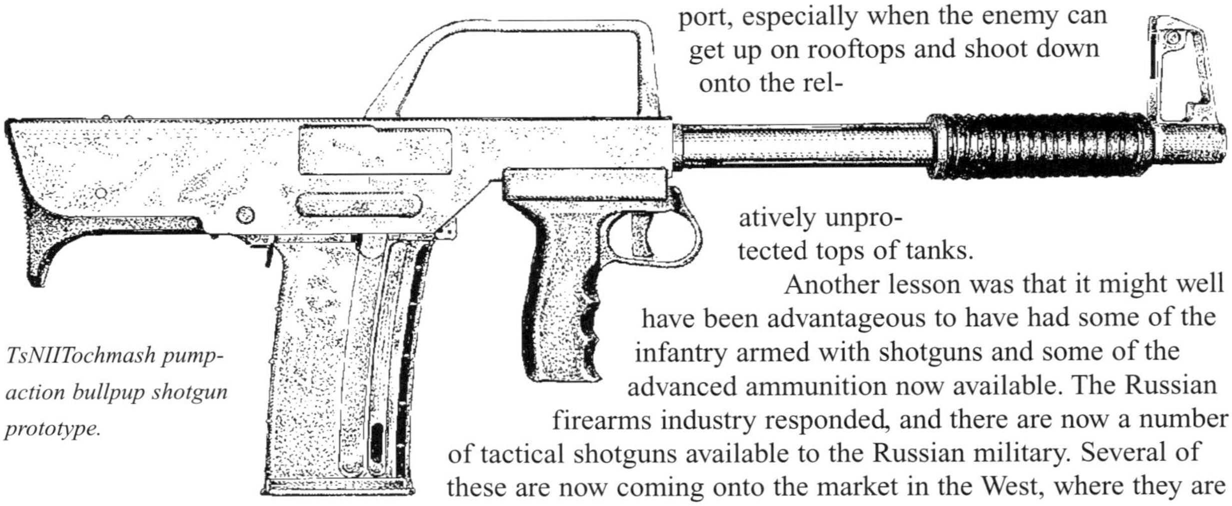

TsNIITochmash pump-action bullpup shotgun prototype.

Another lesson was that it might well have been advantageous to have had some of the infantry armed with shotguns and some of the advanced ammunition now available. The Russian firearms industry responded, and there are now a number of tactical shotguns available to the Russian military. Several of these are now coming onto the market in the West, where they are relatively unrestricted because they are shotguns, which Western politicians still view as relatively benign firearms. All of the Russian "big three" firearms manufacturers have entered the tactical shotgun market, and as are many of Russia's other firearms covered here, Russian tactical shotguns are in many instances very innovative and technically advanced, as we shall see.

KS-23/KS-23M "SPECIAL CARBINE"

If sheer size can be considered an innovation, the TsNIITochmash KS-23 tactical shotgun, called a "special carbine" by the Russians, is the epitome of innovation. It is probably the world's largest production shotgun and arguably the largest pump-action shotgun ever produced. At 23mm-bore diameter, the KS-23 is approximately 6-gauge. It fires a number of different shells, including buckshot, tear gas, rubber bullet, and antivehicular. The antivehicular round is claimed to shatter an engine block upon impact! The recoil of the KS-23 is prodigious, which should come as no surprise. One Western visitor to TsNIITochmash knocked his dental bridgework loose with his first KS-23 shot, and when his Russian hosts insisted that he take a second shot, he fractured his jaw!

The KS-23M is a short-barreled, pistol-grip version of the KS-23, which can be fitted with a wire stock. This weapon has been seen in the hands of SPETsNAZ forces in Russia and in Chechnya. Firing this huge weapon using only a pistol grip is another one of those experiences that separate the true firearms aficionado from the dilettante. What is perhaps even more surprising than a pistol-grip version of the KS-23 is a commercial sporting version, but it does exist. A detachable box magazine version of the KS-23 has been seen from time to time, but TsNIITochmash officials have not responded to queries about this version.

In fact, the KS-23 is a fairly conventional shotgun design save for its size. It is a conventional, underbarrel-tube-magazine-fed, pump-operated shotgun. The bolt is a multilug design that rides in a carrier reminiscent of Stoner systems. According to some Russian sources, the barrel of the KS-23 is rifled. There are

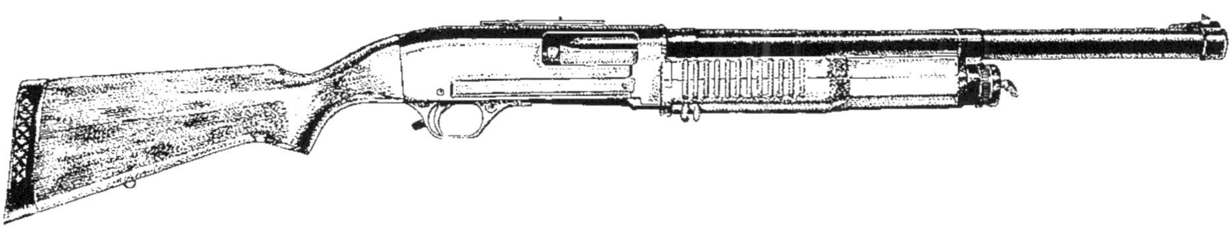

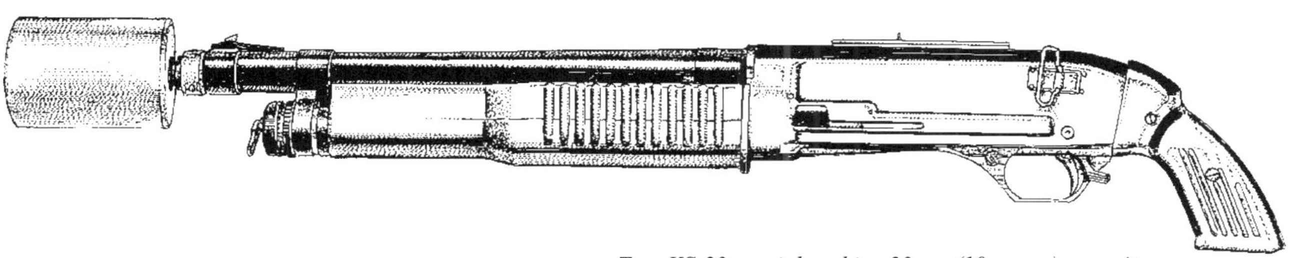

Top: KS-23 special carbine 23mm (10 gauge) capacity.

Above: KS-23 Drozd *(Thrush) shotgun with "cherry" gas projectile (CN gas).*

special muzzle-mounted grenade launchers to fire a variety of grenades. One is a "cup" type of launcher and the other a "spigot" type. The launchers slide over the muzzle and attach via a simple wing nut clamp. The KS-23 and KS-23M are both in service with MVD forces.

KS-23/KS-23M SPECIFICATIONS

Caliber	23mm (6 gauge)
Operation	Manual, Pump Action
Weight	8.5 lbs./7.6 kg. (3.85/3.48 kg.)
Length	41/25.5 in. (1040/650mm)
Barrel Length	19.6/13.7 in. (500/350mm)
Effective Range	100–150 m.
Magazine Capacity	3 Rounds (+1)
Sights	Front, Blade; Rear, Notch

RM/RMB-93/94/96 TACTICAL SHOTGUN

One of the more innovative tactical shotguns from Russia is the Tula KBP-manufactured RM-93 series. The gun has been seen in designations including RM-93, RMB-93, RM-94, and, most recently, at the 1998 SHOT Show in Las Vegas, RM-96. The difference between these guns is difficult to determine. When I asked Tula representatives about the differences, they said that there were no differences between the guns. This, of course, begs the question of why Tula gave them different designations in the first place. Despite this, the guns appear virtually identical, and I will treat them as one for the purposes of our discussion here.

The RM-93 series is very unusual in that the magazine is above the barrel of the gun, and the entire barrel assembly is pumped forward to cycle the action. The notion of pumping forward seems foreign (no pun intended), but in practice it is not. The author has fired the very similar South African Neostead tactical shotgun extensively, and pumping forward to cycle the action takes virtually no getting used to whatsoever. The RM-96 that I handled in Las Vegas in January 1998 was very smooth in operation and fed dummy rounds flawlessly. This does not, of course, mean that the gun will function similarly in the field, but it does give an indication. Shells are loaded into the RM-93 magazine through a loading door in the top of the receiver. The folding stock must be extended to load the gun; this is a disadvantage if the gun is being used tactically with the stock folded. The stock can, however, be removed. To load, the magazine cover is opened to expose the loading port. Part of the feed mechanism is housed in the magazine cover. When the RM-93 is loaded, the cover closed, and the barrel assembly pumped forward, a guide presses the next shell down into position against the breech face as the spent shell is ejected. Like the South African Neostead, the RM-93 has no bolt per se. Once the shell is in position against the breech face, the barrel assembly is pulled to the rear and telescoped over the shell. (This virtually eliminates extraction and ejection problems.) The gun is then ready to fire. When the barrel assembly is pumped forward, the spent shell is held in place against the breech face and ejected straight down as the following shell comes into position for chambering. The trigger of the RM-93 is double action only (i.e., it must go through a long, relatively heavy pull, which cocks and releases the hammer each time the gun is fired). This

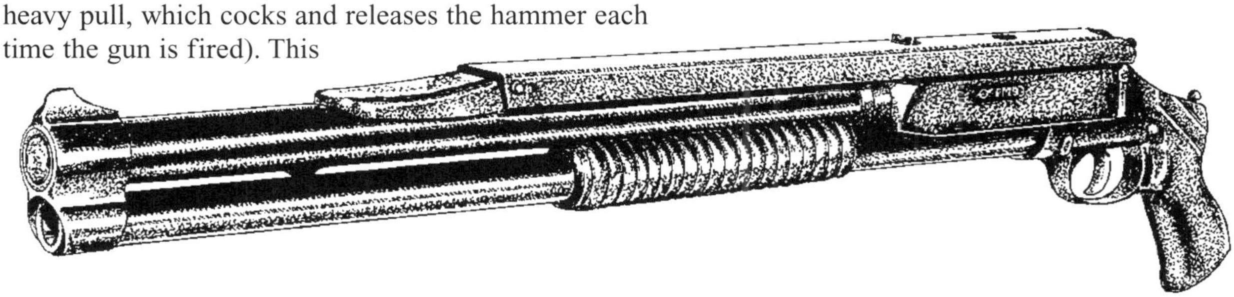

Ruzaje Magazinoje (Magazine Gun, RM) *RM-93. Typical Russian proof marks for a 12-gauge shotgun, Tula Proof House.*

has advantages for today's law enforcement personnel, who face liability problems when using firearms with triggers that are too light, and also allows a second strike on a misfired cartridge, although there are those who would argue that the best tactical course of action is to immediately cycle the gun, dump the misfire, and get a fresh shell in the chamber. It is not the purpose of this book to argue CQB tactics; the fact is that the manufacturer of the RM-93 states that a "second strike" capability was one of its performance goals. Perhaps the MVD desired this feature.

The RM-93 is available in a variety of barrel lengths and finishes, including stainless steel. Chokes are replaceable and are held in place by a screw-on muzzle clamp. Controls are completely ambidextrous. The RM-93 is chambered for 70mm (2.75in) shells. It is also available in sporting versions in varying types of finish and accessories.

RM-93 SERIES SPECIFICATIONS

Caliber	12 Gauge, 2.75 in. (70mm) Shells
Operation	Manual, Pump Action
Weight	5.5 lbs./7.6 kg. (2.5 kg.)
Length, Stock Extended	36.2 in. (920mm)*
Stock Folded	26.3 in. (670mm)*
Barrel Length	21.3 (542mm)**
Effective Range	25-50 m.
Magazine Capacity	7 Rounds
Sights	Front, Blade; Rear, Notch

* Overall length depends on barrel length. Lengths given are for guns with standard 21.3-inch barrels.
** Standard barrel length.

SAIGA (STEPPE ANTELOPE) 12 AND *SAIGA* 20 TACTICAL SHOTGUNS

The *Saiga* series of tactical shotguns is fairly well known by dint of the fact that Izmash has sold some numbers of its *Saiga*-410 shotgun in the West, particularly in the United States. The *Saiga* 410 series shotguns are not included here because they cannot be considered truly tactical shotguns due to their limited ballistics. On the other hand, both the *Saiga*-12 and *Saiga*-20 guns have tactical capabilities. Because of their similarities, we shall consider them together.

The transition from AK rifle to *Saiga* shotgun was not an easy one; there was a great deal of engineering modification necessary to attain the legendary Kalashnikov reliability. Interestingly, the chief engineer on the *Saiga* project was Gennady Nikonov, designer of the AN-94 assault rifle described above. Besides having to enlarge the receiver to accommodate the larger shells, it was necessary to move the guide rails on which the bolt rides to the bottom of the receiver. The ejection port also had to be dramatically enlarged, resulting in an opening behind the bolt. It was therefore necessary to include a sliding "shutter," which is attached to and moves with the recoil mechanism to close the rear portion of the ejection port

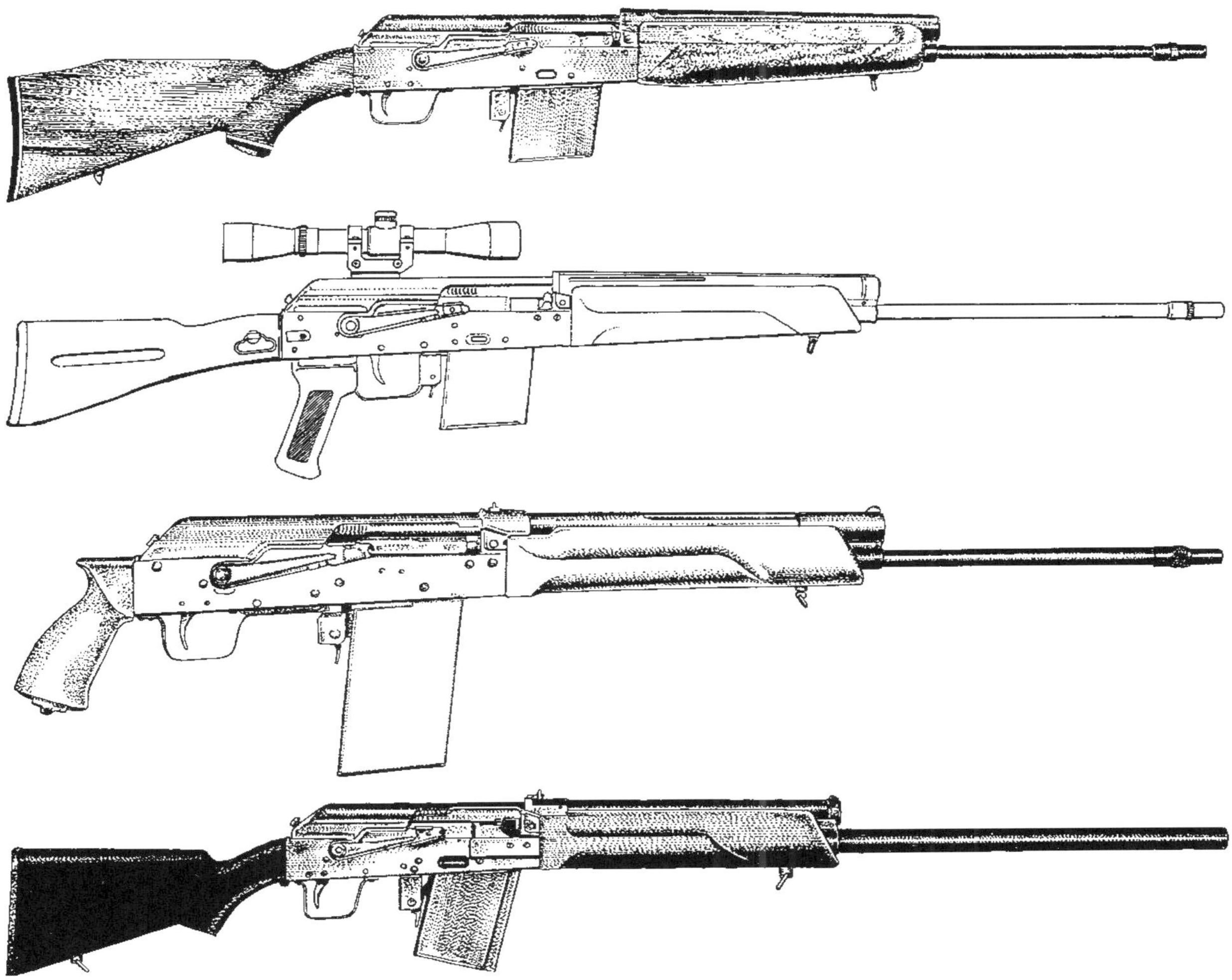

Top: Saiga-*410 Magnum with intermediate barrel.*

Upper Middle: Saiga-*410S with long barrel.*

Lower Middle: Saiga-*410 Magnum with pistol grip and 10-round magazine.*

Bottom: Saiga *20-gauge in 70mm and 76mm.*

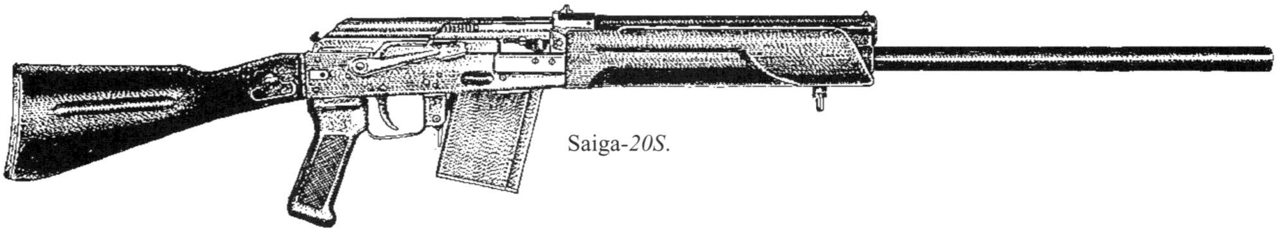

Saiga-*20S.*

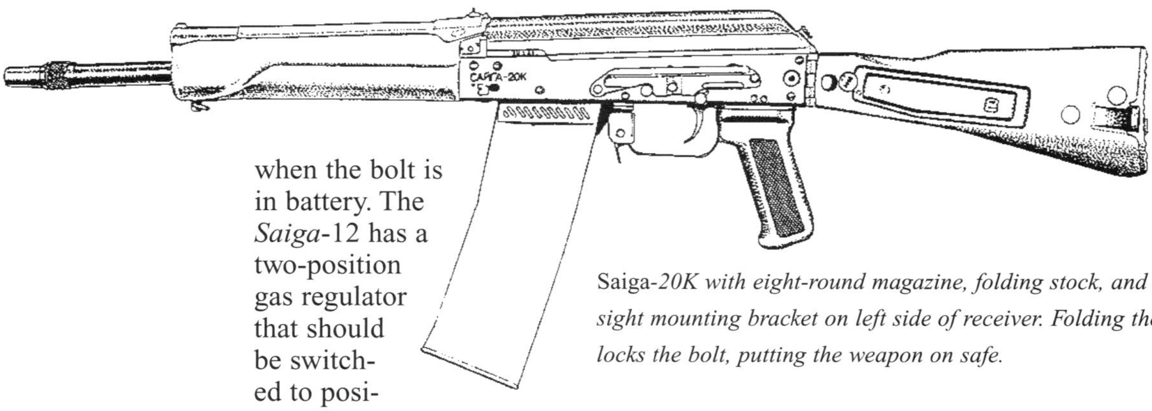

Saiga-*20K with eight-round magazine, folding stock, and optical sight mounting bracket on left side of receiver. Folding the stock locks the bolt, putting the weapon on safe.*

when the bolt is in battery. The *Saiga*-12 has a two-position gas regulator that should be switched to position 2 when using three-inch magnum shells. Although the gun will function with the selector in the "low" position, the manufacturer recommends that high-velocity shells not be extensively fired through the gun with the selector in that position. To do so could result in reduced service life.

The *Saiga* guns feed from a black, glass-fiber-reinforced polymer magazine. Magazines are available in either five- or seven-round capacities. An experimental 10-round sheet steel magazine has been successfully tested and produced in limited quantities. Choke is adjustable via inserts that are retained by a screw-on retaining collar. The stock and forearm are essentially similar to those of the AK-100 series rifles (see Chapter 1). The stock folds to the left on tactical versions of the Saiga. Alternatively, the stock may be removed and a pistol grip affixed in its place. According to Izmash sources, the *Saiga* shotguns are in full production, but the extent of their adoption by the Russian military and police is unknown.

SAIGA-12S/12K TACTICAL SHOTGUN SPECIFICATIONS

Caliber	12 Gauge, 2.75 in. or 3-in. Shells
Operation	Gas, Semiautomatic
Weight	7.92/7.7 lbs. (3.6/3.5 kg.)
Length, Stock Extended	41.7/35.8 in. (1,060/910mm)
Stock Folded	32.2/26.3 in. (820/670mm)
Barrel Length	22.8/16.9 in. (580/430mm)
Effective Range	25-50 m. (Auto Dependent)
Magazine Capacity	5 or 7 Rounds
Sights	Front, Brass Bead; Rear, Notch

CHAPTER FIVE

GRENADE LAUNCHERS

Since they were introduced in the mid-1960s, grenade launchers have been an important and integral part of the basic arsenal of the infantryman. They began with stand-alone launchers, such as the U.S. M-79, and eventually evolved into launchers mounted beneath the barrel of the rifle, which is the basic soldier's weapon. In most armies today, there are one or two grenadiers in every infantry squad. Grenade launchers also developed into fully automatic weapons-automatic grenade launchers, also known as grenade machine guns. They have taken other forms as well, and the Russians seem to have tried almost all of them.

In this chapter we will explore Soviet and Russian grenade launchers of all types. I will be departing somewhat from the criteria for inclusion that have formed the basis for all previous chapters in that I am including a few weapons that have been widely fielded in the Russian military because they are relatively unknown here in the West.

The Russians lagged behind the West in their efforts to develop grenade launchers, but it did not take them long to not only catch up, but in many areas surpass their Western rivals, as we shall soon see. The Russians have developed a variety of specialized grenade launchers, ranging from providing common land defense to defending ships and port facilities against underwater swimmers. They have even developed silent grenade launchers! The reader will note the absence of the AGS-17. This early automatic grenade launcher was purposely omitted because it is so well known and there is a wealth of data available for it.

AGS-30 AUTOMATIC GRENADE LAUNCHER

The Russian AGS-17 was introduced into service in 1975 and has proven to be reliable and effective in service. But like so many automatic grenade launchers of its era, the AGS-17 is heavy. The launcher alone weighs 39.6 pounds (18 kilograms), and the mount which is necessary to fire the weapon adds an additional 26.4l pounds (12 kilograms), bringing the total empty weight to 66 pounds (30 kilograms). Although the weight of the AGS-17 is somewhat less than the Saco Defense Mark 19, for example, it still is a heavy system for the infantry to have to carry outside of a vehicle. The newly adopted AGS-30 apparently solves this problem for the Russians.

The AGS-30 first appeared in 1994 as a prototype carrying the designation TKB-722K. This prototype has apparently since passed its troop trials and been adopted by the Russian military. Versions of the AGS-30 now appearing in Russian sales literature show better finish than the early versions. Generally, the AGS-30 performs the same mission as the AGS-17 but is half the weight of the older system while firing the same ammunition and using the same sights and accessories. The weight of the system has been decreased sufficiently to allow reduction of the crew from three to two men. Rate of fire of the AGS-30 is slightly higher than that of its predecessor, but maximum and effective ranges are identical. Like the

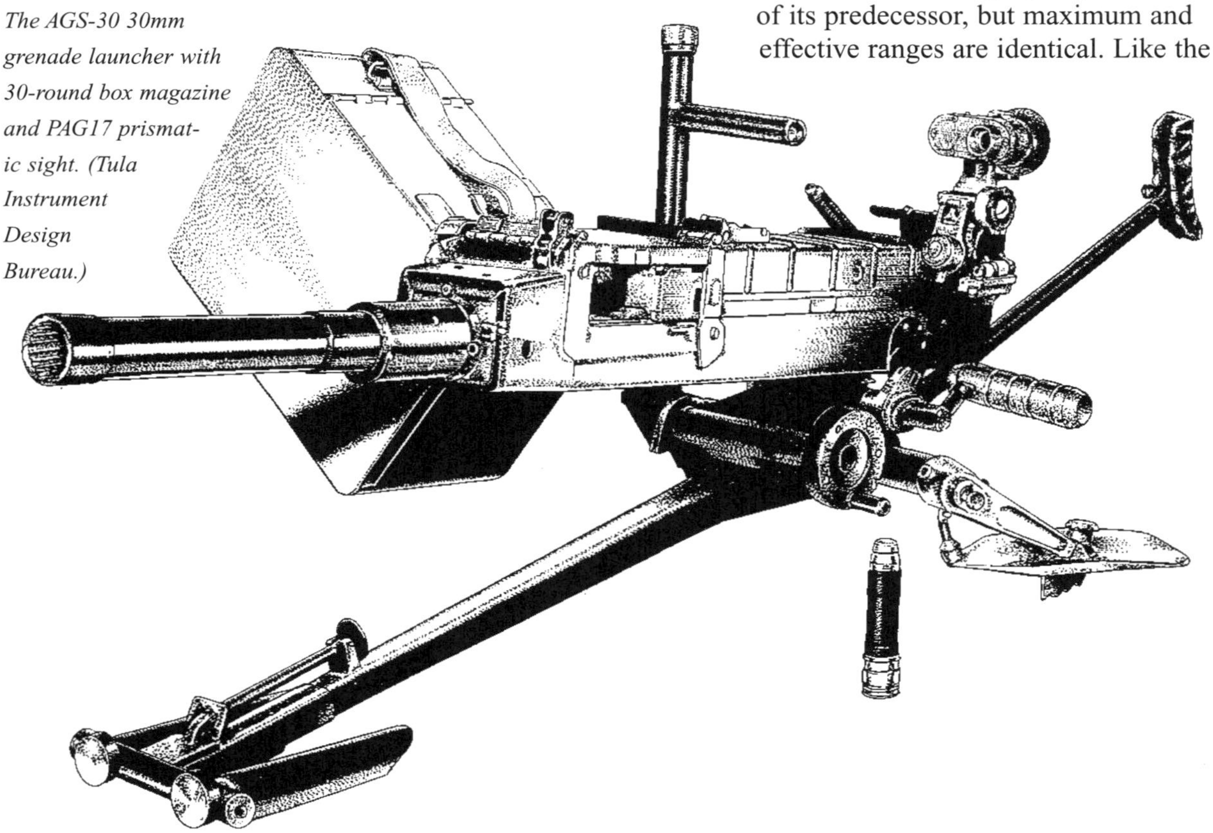

The AGS-30 30mm grenade launcher with 30-round box magazine and PAG17 prismatic sight. (Tula Instrument Design Bureau.)

AGS-17, the AGS-30 is claimed to be blowback operated, but blowback operation requires a very heavy bolt and recoil mechanism, so it is suspected that the AGS-30 may be delayed blowback, as this method would allow a reduction in weight by reducing the mass of the bolt. Sales literature suggests that the AGS-30 also has a buffer system to reduce recoil, but this is not definitely stated. Nonetheless, given the fact that Tula KBP, which manufactures the AGS-30, has engineered out half the weight, while retaining the same VOG-17 cartridge, strongly suggests that a buffering system will be necessary to cope with the strong recoil forces generated as the AGS-30 is fired.

The AGS-30 appears to be a strong contender in the international market if it delivers performance as advertised. The weapon is by far the lightest automatic grenade launcher in the world, and this can only win the hearts of the infantrymen who must carry it into battle, if it is as effective as its predecessor.

AGS-30 SPECIFICATIONS

Caliber	30x29mm
Operation	Blowback, Select Fire
Weight (Launcher and Mount)	35.2 lbs. (16 kg.)
Length	43.3 in. (1,100mm)
Barrel Length	App. 14 in. (355mm)
Effective Range	Up to 1,700 m.
Magazine Capacity	30 Rounds (Belt Fed from Box)
Sights	Optical, PAG-17; Mechanical: Front, Post; Rear, Notch

DP-61 "ANTISABOTEUR ROCKET GRENADE LAUNCHER"

The DP-61 and its companion weapon that follows, the DP-64, are both intended for the same mission, but are somewhat different in execution and function, hence we shall treat them separately. Both are intended to protect ships, ports, and coastal installations from combat swimmers and to indicate their location if they are not destroyed by the explosive grenade.

The DP-61 is a simple, single-barreled launcher that fires a 55mm rocket grenade to a maximum range of 500 meters. The GR-55M is a high-explosive grenade, and the GRS-55 is used for marking the location of swimmers. Since the grenade is rocket propelled, it probably cannot be fired from enclosed spaces. The grenade fired by the DP-61 is slightly larger and has a slightly greater effective kill radius than the DP-64, but the differences are slight. It is curious that Tula would develop two virtually identical weapons, but it has. Other than the information contained here, there is little else known about the DP-61, other than the fact that it is in service with the Russian Defense Ministry, probably the navy.

DP-61 ANTISABOTEUR ROCKET GRENADE LAUNCHER SPECIFICATIONS

Caliber	55mm
Weight	13.2 lbs. (6 kg.)
Effective Range	500 m.
Effective Radius	18 m.

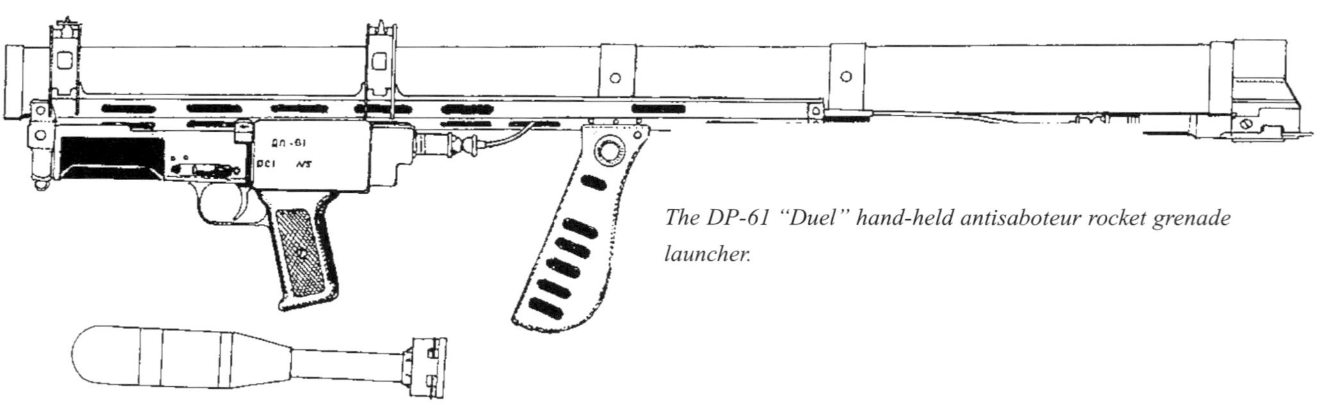

The DP-61 "Duel" hand-held antisaboteur rocket grenade launcher.

DP-64 NEPRYADVA ANTISABOTEUR GRENADE LAUNCHER

The DP-64, like the DP-61 above, is intended to defend ports, ships, and coastal installations against combat swimmers. The DP-64 also fires two types of grenades, the FG-45 high-explosive grenade and the SG-45 marking grenade. The FG-45 grenade has a pressure fuse that is set prior to firing. Once it reaches the preset depth, it explodes the grenade, which achieves its effects via overpressure rather than fragmentation. The SG-45 grenade is equipped with a contact fuse. When the grenade explodes, it sends a flare to the surface, which burns brightly for 50 seconds to indicate the approximate location of the enemy diver. How the grenade is prevented from exploding when it strikes bottom is not made clear in any of the available Russian literature. Unlike the DP-61, the DP-64 is a closed-breech weapon that can be fired from an enclosed space, which the Russians make clear in their sales literature regarding the system. The DP-64 has two vertically arranged barrels.

DP-64 NEPRYADVA ANTISABOTEUR GRENADE LAUNCHER SPECIFICATIONS

Caliber	45mm
Launcher Weight	22 lbs. (10 kg.)
Grenade Weight	1.43 lbs. (.65 kg.)
Effective Range	400 m.
Effective Depth	40 m.
Lethal Radius	14 m.

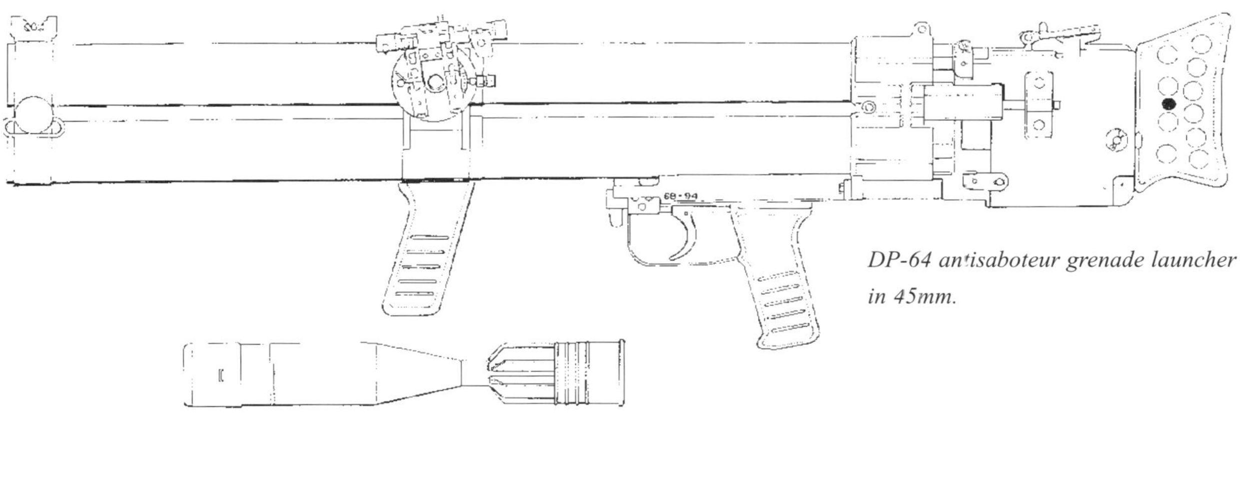

DP-64 antisaboteur grenade launcher in 45mm.

GP-25/GP-30 UNDERBARREL GRENADE LAUNCHERS

The genesis of the GP-25/30 goes back to the days of the Afghan War when the earliest versions were used by the Soviet forces against the Mujahideen. The GP-25 and its simplified follow-on, the GP-30, are simpler in operation than the U.S. M203. The VOG-25 and VOG-25P grenades are simply loaded from the muzzle and fired. Both grenades are high explosive with prefragmented steel liners; the VOG-25 has an impact fuse, and the VOG-25P is a "bounding" round that hits the ground and then is propelled up approximately 1.5 meters before it explodes, spraying fragments over a lethal radius of approximately 6 meters. Both self-destruct in 14 to 19 seconds if they do not explode on impact. The VOG-25 has a rounded ogive, while the VOG-25P has a conical ogive.

The two grenade launchers are, for practical purposes, identical, but there are some differences. According to manufacturer's literature from Tula KBP, the GP-30 is lighter and shorter, making the weapon to which it is attached easier to handle. It is also less expensive to manufacture than the GP-25. More important, unlike the GP-25, the GP-30 can be installed on virtually any rifle, not just AK-type weapons. Other than these differences, the two launchers are the same, with a steel rifled barrel, double-action-only trigger mechanism, and sights graduated to 400 meters.

Top: GP-25 underbarrel 40mm grenade launcher.

Bottom: AKS-74U with GP-25 Kastyor grenade launcher. The GP-25 is attached to a barrel extension tube replacing the muzzle flash hider. The standard rubber pad used with metal stocked weapons is fitted. Note the mounting plate on the receiver nor an optical sight.

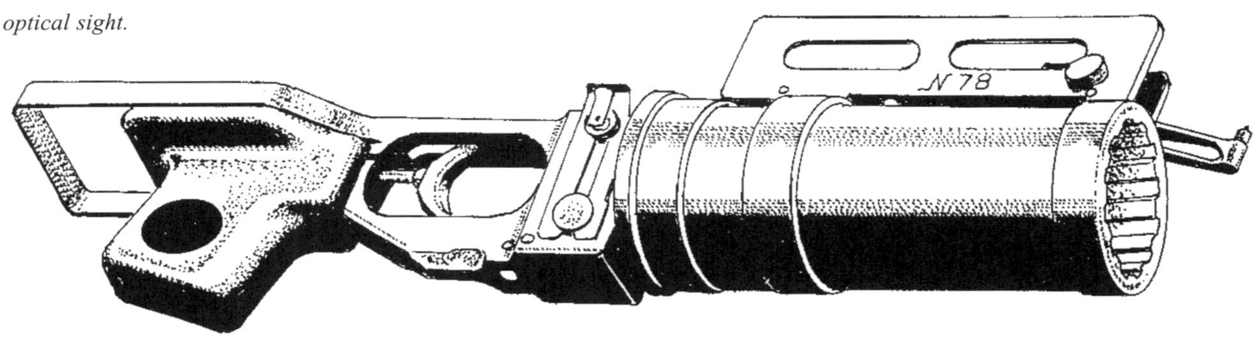

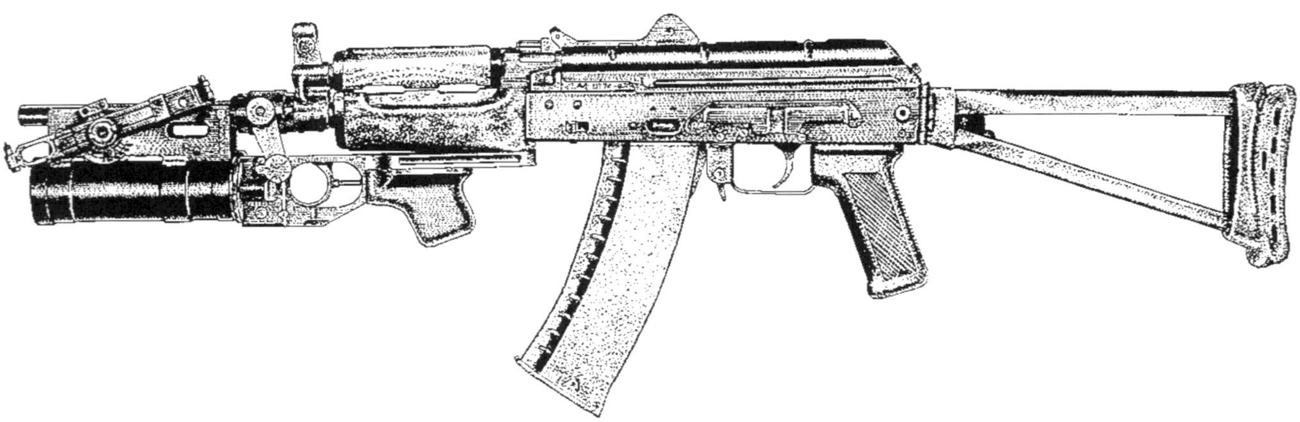

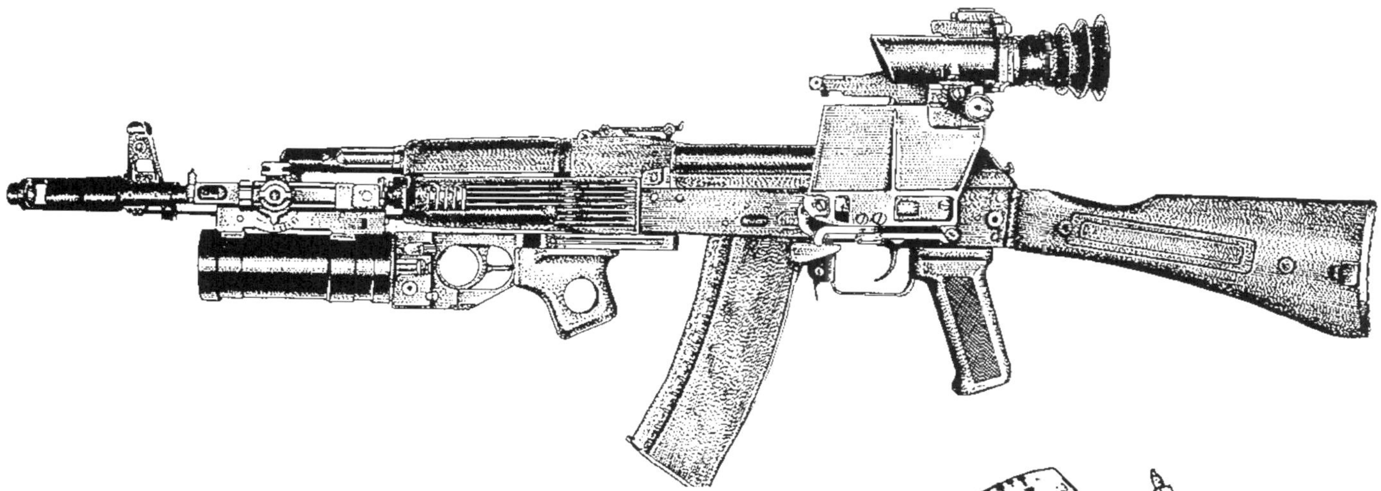

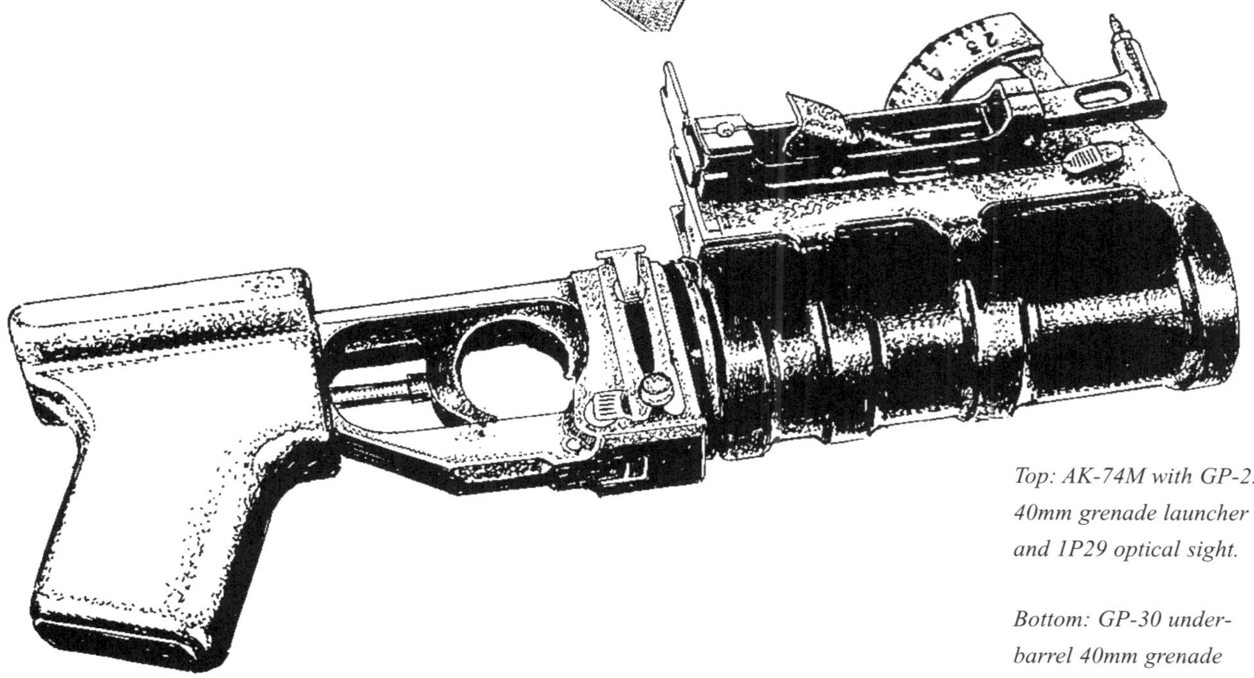

Top: AK-74M with GP-25 40mm grenade launcher and 1P29 optical sight.

Bottom: GP-30 under-barrel 40mm grenade launcher. Note that the GP-30 is shorter than the GP-25, there is no frame at the rear, and the sight is mounted on the right side.

GP-25/GP-30 SPECIFICATIONS

Caliber	40mm
Weight	3.32/2.8 lbs. (1.5/1.3 kg.)
Length	12.7/10.8 in. (323/276mm)
Effective Range	200–400 m.
Rate of Fire	5 RPM

GP-95/GP-97 GRENADE LAUNCHER

The GP-95 is a compact grenade launcher developed for mounting on the A-91 series of compact assault rifles manufactured by Tula KBP. Compared to the earlier GP-25/30 launchers, the GP-95 weighs approximately the same and has the same capabilities. The GP-97 appears to be identical to the GP-95, save for the fact that it has been modified to mount on top of the barrel of the A-91M bullpup assault rifle (see Chapter 1), rather than underneath. Both grenade launchers fire standard VOG-25 and VOG-25P rounds.

GP-95 SPECIFICATIONS

Caliber	40mm
Weight	3.3 lbs. (1.5 kg.)
Length	7.2 in. (185mm)
Effective Range	200–400 m.
Rate of Fire	5 RPM

RG-6 GRENADE LAUNCHER

The RG-6 is a development of Tula KBP and is still in prototype form. Whether it will ever be fielded is questionable at the moment, considering some teething problems that have been discussed in the Russian military press.

The idea for the RG-6 came from the South African Striker 12 revolving shotgun and similar designs dating to the 1930s. The Russian designers intended to produce a weapon that was more transportable than an automatic grenade launcher but capable of launching several grenades. The RG-6 is the brainchild of the prolific Valery Telesh and colleague B.A. Borozov. These gentlemen worked night and day to design their weapon and went from emergency MVD requirements to working prototype in only 11 days! The requirement was received on 10 November 1993, the first prototype was prepared by 21 November, and six prototypes were in the hands of the customer for test and evaluation during the first quarter of 1994.

The RG-6 is essentially six GP-25 barrels mounted to a central rotating axis. The six rotating barrels are each rifled, while the main barrel is not. When the trigger is squeezed, a spring pushes the barrel group forward against the main barrel to form a tight seal. Once the round is launched, the cylinder retracts and is automatically rotated one-sixth turn, bringing a fresh round into position. Apparently, the spring that operates the cylinder must be wound like that of a clock.

As mentioned, there are some teething problems with the RG-6, which is the norm with any new weapon. First, it doesn't always fire all six grenades, leaving one or two in the cylinder to perhaps be unloaded. Unloading may be problematic, because the RG-6 uses standard VOG-25 and VOG-25P rounds, which are muzzle loaded and apparently a snug fit in the barrel's rifling. The RG-6 is also slow to load. As one tester pointed out, the RG-6 fires its six rounds more rapidly than a GP-25, but the GP-25 can be fired six times while the RG-6 is being reloaded. Also, parts of the RG-6 came loose during the tests and caused problems of reassembly after reloading the weapon. The conclusion of the operators who tested the RG-6 was that it needed work and a better range of ammunition, including smoke, illumination, shaped charge, and flechette, if it was going to fulfill its potential. They did not believe the RG-6 to be suited for general issue to troops, but stated that it would be very useful in MOUT and special operations. In fact, two RG-6 grenade launchers were used in Chechnya.

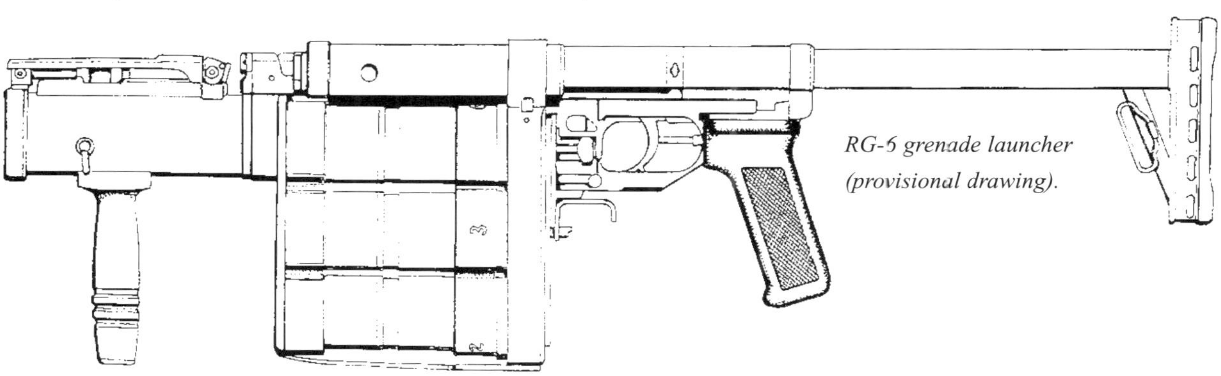

RG-6 grenade launcher (provisional drawing).

It should be noted that the latest Tula KBP literature on the RG-6 states that it can be easily unloaded by simply pressing an extractor, so the company is obviously working on correcting defects in the weapon. The RG-6 may yet find a home in the Russian military.

RG-6 GRENADE LAUNCHER SPECIFICATIONS

Caliber	40mm
Weight	12.7 lbs. (5.8 kg.)
Length, Stock Extended	26.7 in. (680mm)
Stock Folded	20 in. (510mm)
Effective Range	200–400 m.
Rate of Fire	12–15 RPM

GM-94 MAGAZINE GRENADE LAUNCHER

Another development of Tula KBP, the GM-94 was derived from the earlier GM-93, which was apparently a developmental launcher. The GM-94 fires a new 43mm grenade similar to the Western 40mm tube-launched grenades. Other than the fact that the launcher fires a new type of round, little is known of the ammunition other than the fact that it is issued in several types. Unlike 40mm VOG-25 ammunition, which as far as is known comes in only two versions, the new 43mm VGM ammunition can be had in high explosive, thermobaric (a special, very potent fuel-air explosive), shaped charge, illuminating, gas, signal marker, rubber baton, and flash-bang. It seems that Tula has developed something for everyone who might consider adopting the GM-94!

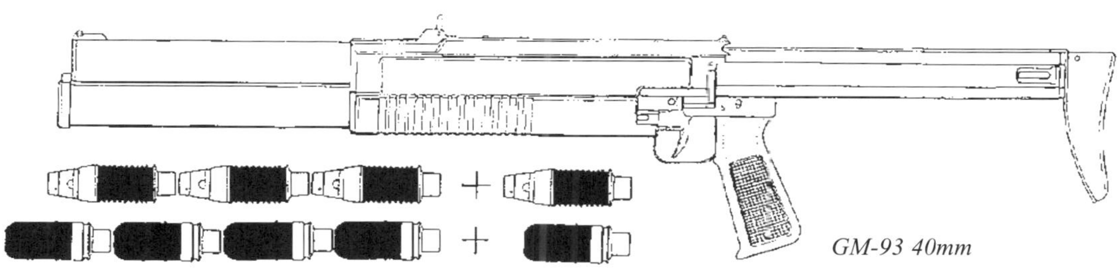

GM-93 40mm magazine-fed grenade launcher.

The GM-94 is obviously derived from the RM-93 shotgun or vice versa. At any rate, the magazine is above the rifled barrel, which is pumped forward to cycle the launcher's action. The "straight line" design of the barrel in relation to the stock helps reduce felt recoil, which is essential in a weapon such as this, which surely has heavy recoil. The magazine holds three rounds of all ammunition types except the rubber baton type, of which four can fit into the magazine. An extra round can be placed in the chamber, adding one round to the total load. Unlike with the RG-6, loading the GM-94 is apparently simple. To reload, one merely lifts the loading cover like that on the RM-93 shotgun and shoves in the fresh rounds. This also allows for quick tactical reloads of different types of ammunition should the need arise. Also, if a round misfires, all the operator has to do is cycle the action to eject the dud and chamber a fresh grenade.

According to the manufacturer, the GM-94 outperforms any hand-held grenade launcher in the world. That is a strong claim, but the GM-94 appears to be a well-thought-out design for a magazine-fed, hand-held grenade launcher. Whether it will find a home in any of the world's military forces remains to be seen. Most military forces prefer not to have a dedicated grenadier (grenadier with only a grenade launcher and no rifle) for a number of tactical reasons, but the portable firepower of the GM-94, along with its versatile array of ammunition, may well make many friends for this unusual magazine-fed, pump-action grenade launcher.

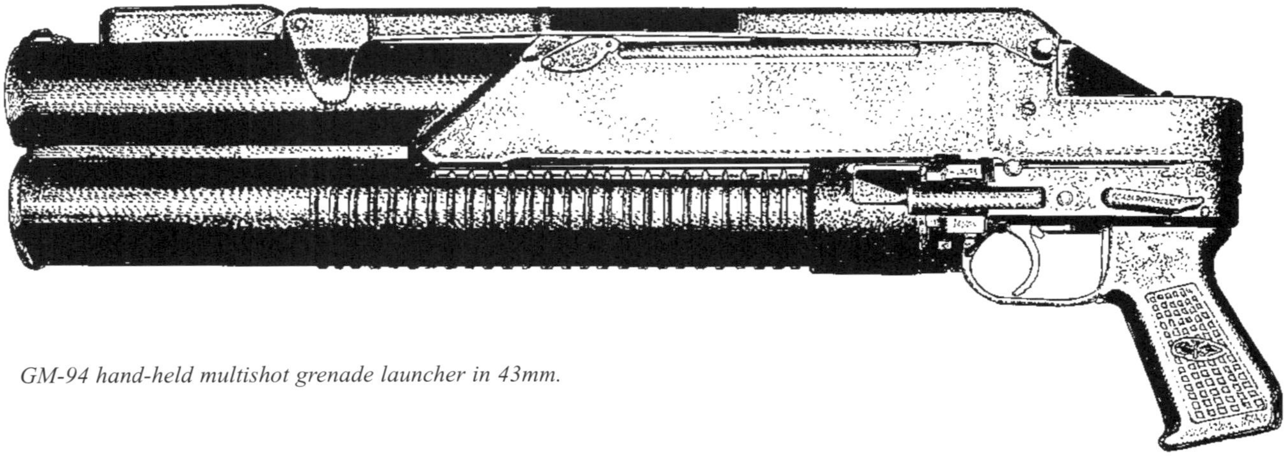

GM-94 hand-held multishot grenade launcher in 43mm.

GM-94 PUMP-ACTION GRENADE LAUNCHER SPECIFICATIONS

Caliber	43mm
Weight	10.6 lbs. (4.8 kg.)
Length, Stock Extended	29.2 in. (742mm)
Stock Folded	20.7 in. (528mm)
Barrel Length	App. 16 in. (406mm)
Effective Range	75–300 m. (Ammunition Dependent)
Magazine Capacity	3 or 4 (+1) Rounds
Sights	Front, Blade; Rear, Notch

BS-1/RGA-86 SILENT GRENADE LAUNCHER

Of all the innovations that Russian firearms designers have engineered, the silent grenade launcher must take a place of honor alongside underwater firearms and semiautomatic pistols using silent gas-trap cartridges. We will not enter into the tactical efficacy of these weapons from a Western viewpoint, but it must be remembered that most Russian firearms are designed to a government requirement of some sort. The BS-1/RGA-86 has been known for some time in Western intelligence circles but only recently has come to public light via the Russian press. When it first became known in the West, it was referred to as the RGA-86, but recent Russian press articles refer to it as the BS-1, and I will use that designation. Not surprisingly, little concrete data are available on this mysterious grenade launcher, but from what is available much can be deduced.

First, the BS-1 grenade cartridges almost certainly employ a gas-trap piston mechanism of some sort to propel the grenades. In fact, the BS-1 grenade launcher may even use the same SP-4 cartridge as the PSS-1 silent pistol (see Chapter 2), but the few photographs that show the grenade cartridges are not sufficiently clear to make a determination. There are two possible ways the BS-1 can function.

The cartridge shown in the photos may be a typical grenade cartridge with a crimp, leading to the conclusion that the silencing mechanism is in the launcher itself in the portion behind the chamber. In this case, there would be a gas-trap mechanism behind the chamber containing a piston that drives the grenade. The grenades shown have no propellant attached, so the propelling force must somehow be transferred from the launch cartridge through the portion of the launcher between the cartridge chamber and the grenade launch tube. Moreover, there are slots in the breech end of the launch tube through which the grenade body shows when the BS-1 is loaded. This leads to the conclusion that there is little or no gas pressure inside the launcher barrel. Therefore, if this method of propelling the grenade is used, there must be a piston-type mechanism in the portion of the launcher behind the barrel that propels the grenade out of the barrel and downrange. Further, the portion of the launcher behind the barrel takes up nearly half the launcher, so it is almost a certainty that the suppression mechanism would be contained therein, if this is the operating system used by the BS-1.

A second possible

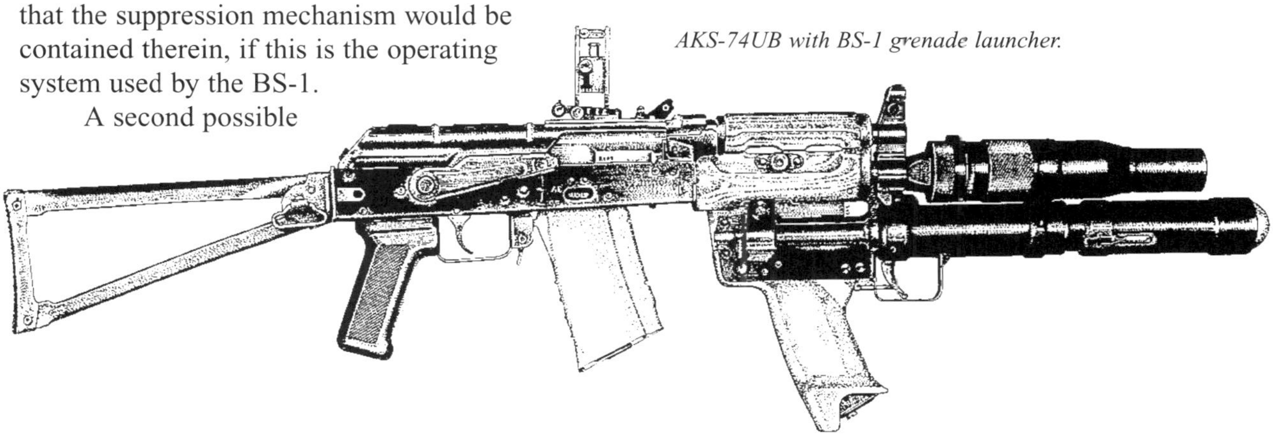

AKS-74UB with BS-1 grenade launcher.

operating system is one in which the BS-1 actually uses the SP-4 cartridge and fires a bullet into the base of the grenade, propelling it out the barrel of the launcher. Either this or the method described above would work, although the "bullet" method probably is simpler. The truth will have to wait until more details are released by the Russian manufacturers of the BS-1.

The BS-1 fires a unique 30mm grenade that is apparently available only with a shaped-charge warhead. Russian documents available to me list only a shaped-charge warhead for the grenade; this is rather surprising, because most other Russian projected grenades have more than one type of warhead. Although the mission profile of this unique grenade-launching system is not known, it clearly is intended for special-operations use, as is the modified AKS-74U, which mounts the BS-1. Given the launch mechanism, the range of the BS-1 must be appreciably less than conventional Russian launchers such as the GP-25/30 and GP-95/97.

Finally, the BS-1 silent grenade launcher feeds its launching cartridges from a detachable box magazine in the grip of the launcher, indicating that it is designed for repeat fire. Whether it is manually operated or semiautomatic, or whether the magazine simply holds extra cartridges is not known. There is no visible ejection port on the launcher, although it is possible that the entire bolt mechanism is blowback operated and cycles to the rear when the grenade launcher is fired. There appears to be a space between the front of the AKS-74U magazine and the rear of the launcher for the bolt to cycle.

In sum, as of the time of this writing, there are many unanswered questions about the BS-1 silent grenade launcher. The weapon is apparently in use by Russian special forces; examples seen in photos have had serial numbers running up into the hundreds, indicating that it is more than just a prototype. The weapon is also probably manufactured by Izmash, as it appears in the book *Kalashnikov Arms*, which is devoted almost exclusively to Kalashnikov firearms produced at Izmash. Even with the few facts I have stated here, much about this weapon remains unknown. There is no specifications table because all known data on this weapon are contained in the text.

CHAPTER SIX

AMMUNITION

Without ammunition, a firearm is nothing more than a very expensive club—and in many instances, a very inefficient club at that. A firearm must be thought of as a system comprising two components: the projectile launcher (or the firearm itself) and the cartridge, which contains the projectile and the necessary components to propel it from the launcher. There are subsystems to both components. For example, the firearm comprises a barrel, receiver, trigger mechanism, and sights. Each of these in turn also comprises subsystems, all of which affect overall external ballistics, including accuracy. The cartridge component of the system also comprises subsystems, including projectile, case, propellant, and primer. These too have an effect on both ballistics and accuracy. As can readily be imagined, the overall design of a firearm to a requirement for the military is a complex process involving many engineering disciplines.

Heretofore, we have explored only one aspect of the overall firearms system—the firearm itself. In this chapter I will discuss the specialized ammunition that the Russian firearms industry has developed to optimize the efficacy of their new firearms.

The reader will not find standard ammunition listed in this chapter. The cartridges discussed here are specialized, and new cartridges, for the most part, were developed especially for the weapons covered in the first five chapters.

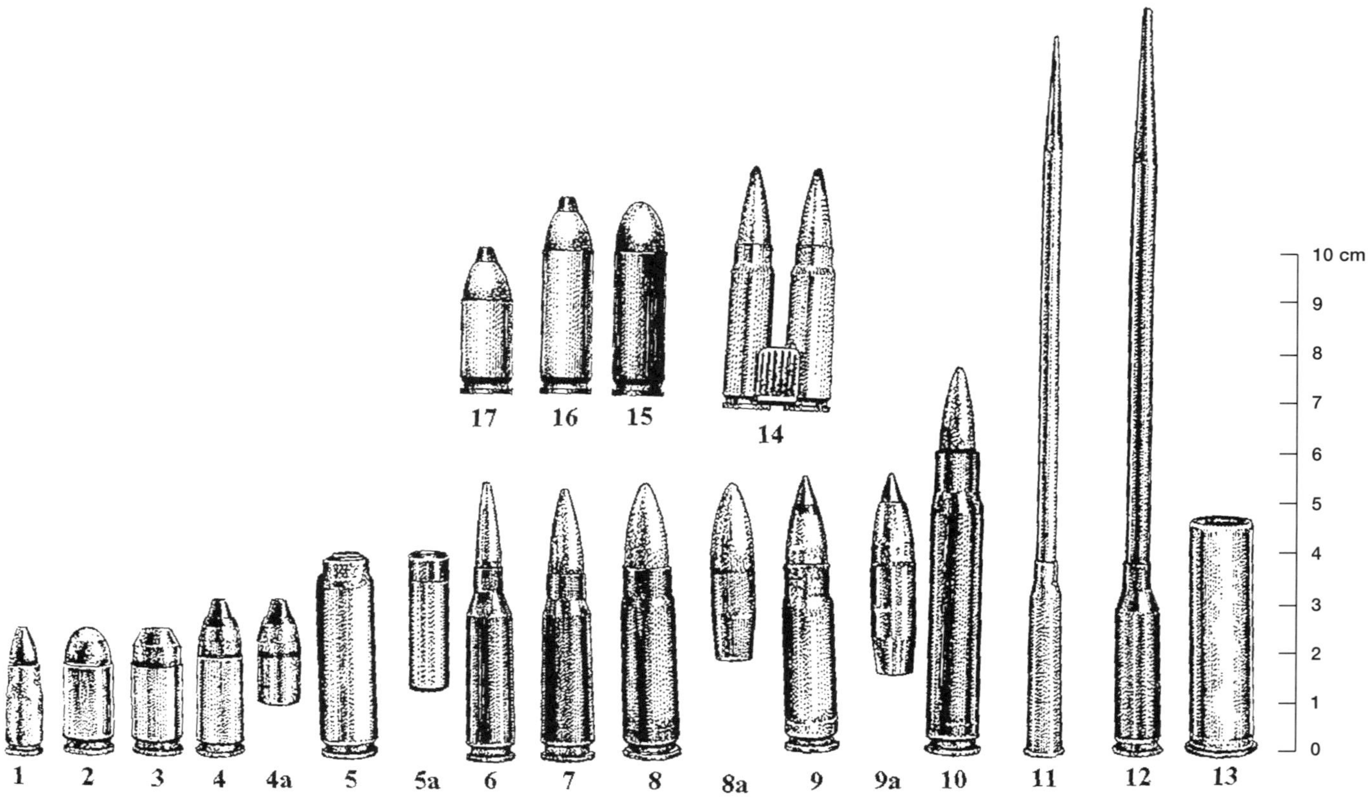

Modern Russian small-arms ammunition:

1.	*5.45x18mm*	*(PSM) Drotik*
2.	*9x18mm 57-N-181S*	*(PM, Klin, Kedr, nn-90, nn93, Bizon)*
3.	*9x18mm 57-N-181SM*	*(PMM)*
4.	*9x21mm RG-054*	*(Gurza)*
4a.	*9mm AP round*	*(Gurza)*
5.	*7.62x41mm SP-4*	*(PSS)*
5a.	*7.62mm steel alloy round*	*(PSS)*
6.	*5.45x40mm*	*(AK-105)*
7.	*7.62x39mm 57-N-321*	*(A-91, AK-103)*
8.	*9x39mm SP-5*	*(9A-91, VSS, OTs-14)*
8a.	*9mm subsonic round*	*(SP-5)*
9.	*9x39mm SP-6*	*(Vikhr, MA-9, AS, OTs-14)*
9a.	*9mm AP subsonic round*	*(SP-6)*
10.	*7.62x62, 8mm SP-3*	*(5-4M)*
11.	*4.5x39mm*	*(SPP-1M)*
12.	*5.56x40mm*	*(APS)*
13.	*12.3x50mm solid shot*	*(Udar)*
14.	*7.62x35mm SP-2*	*(MSP)*
15.	*9x30 GROM*	*(Gepard)*
16.	*9x30 MAC GROM*	*(Gepard)*
17.	*9x19 H.A.C. RG-057 (PP.90M)*	*(PP-90M1)*

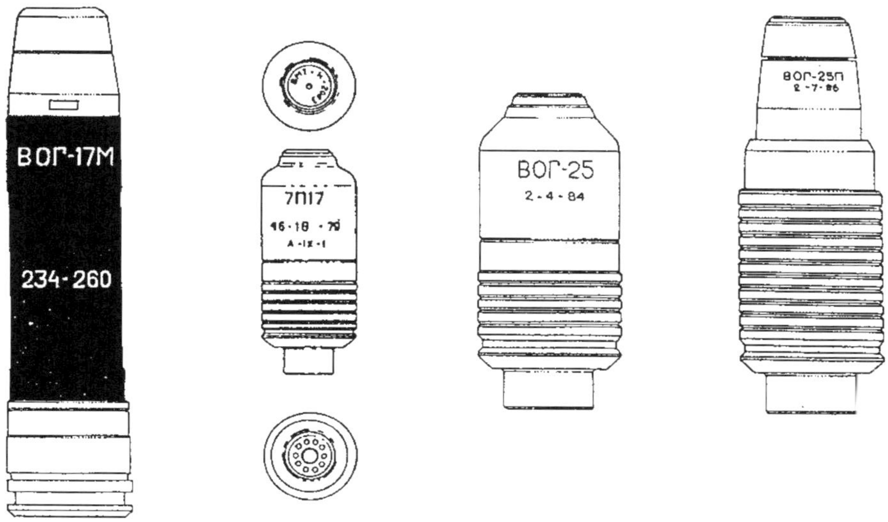

Far left: VOG-17M

Second from left: (VOG-15) 7P17 grenade for GP-15.

Third from left: VOG-25 gunshot from fragmentation grenade-25.

Fourth from left: Gunshot from fragmentation grenade-25 ("jumping"). The extended fuze contains a propelling charge and short relay to hurl the grenade to chest height before it explodes. This results in a smaller explosive charge (VOG-25 contains 48 gms.; VOG-25P 37 gms.).

As is the case with firearms, Russian ammunition producers tend to be secretive with data about their ammunition. For example, they seem reluctant to discuss the exact makeup and internal design of many of the bullets for their new cartridges, stating only that they perform in a certain way, achieving a certain level of penetration against a specific target. The problem with the Russian expression of terminal ballistics is that they seem to have no specific standards, and until very recently the expressions that they use bore no relation to those used in the West. Around 1997, the Russian manufacturers began to use penetration effectiveness against National Institute of Justice (NIJ) targets in some instances. As for exterior ballistics, Russian manufacturers are virtually silent. There is usually nothing given but muzzle velocity and, in some cases, bullet weight. In the cases where those facts are given, exterior ballistics can be extrapolated using computer ballistics programs. Any ballistics data here, aside from velocity and bullet weight, were derived using Blackwell's Load from A Disk ballistic software, which was designed for reloaders but has proven accurate for extrapolating ballistic data for nonstandard cartridges.

The Russians have not only designed new cartridges, but have also improved the ammunition used in earlier weapons, many of which remain in Russian military service. In the latter instance, the Russians have introduced new ammunition with improved penetration capabilities for almost every firearm in their inventory. This includes 5.45x39mm, 7.62x39mm, and 9x18mm calibers. They have also introduced subsonic ammunition in most rifle calibers, including 5.45x39mm, 7.62x39mm, and 7.62x54Rmm. Subsonic ammunition is used with suppressed weapons because with supersonic ammunition there is a loud "crack" as the bullet leaves the muzzle of the weapon and breaks the sound barrier. Subsonic ammunition optimizes the effect of the suppressor. To recover effectiveness lost when the velocity is reduced, the bullet in subsonic ammunition is almost always heavier than that of standard ammunition.

The Russians seem to have concentrated on achieving improved penetration; most of the rifle ammunition manufactured by them since 1991 would be considered armor piercing in the West. The new, improved penetration ammunition is so popular with the Russian military that commanders in Chechnya specifically requested it when they needed ammunition resupply. Why this emphasis on armor-piercing ammunition? The answer is simple. First, the Russian military is very likely to face enemy troops wearing body armor that may defeat the older intermediate-caliber ammunition. Second, Russia faces a crime problem that makes anything in the West pale to insignificance by comparison, and virtually all Russian criminals can be expected to be wearing body armor whenever they encounter police. The police must have ammunition that will defeat both soft body armor and car bodies, hence the development of armor-piercing pistol ammunition. Having laid the general background, we are now ready to explore the specifics of new and improved types of Russian ammunition. All Russian ammunition is Berdan primed.

4.5x39Rmm SPS UNDERWATER PISTOL CARTRIDGE

This highly specialized cartridge has already been briefly discussed under the SPP-1 underwater pistol. The cartridge was designed by the team of Pyotr Sazonov and Oleg Kravchenko, who received a USSR State Prize in 1983 for their work, along with Vladimir Simonov, designer of the SPP-1 pistol.

In their initial experiments, the designers found that standard bullets fired underwater had very limited range and would never work in underwater firearms. The designers therefore began to look at patterning their bullets after the long darts used by divers for spear fishing. This proved to be the solution, and the cartridges for both the SPP-1 pistol and APS underwater assault rifle were designed using the "dart" bullet.

The SPS cartridge fires a 203-grain (13.2-gram) steel bullet at a muzzle velocity of approximately 820 fps (250 mps). The bullet is not spin stabilized; stability is obtained by a cavitation bubble that forms along the interface between the long axis of the bullet and the water through which it passes. According to Russian sources, the bullet does not stabilize at all in open air and has a range of only some 20 meters in that medium, along with degraded accuracy. Effective range of the SPS bullet underwater depends on depth and concomitant water pressure.

SPS cartridges are loaded into the SPP-1 pistol in four-round clips. Basic load for the pistol is four clips.

SPS 4.5x39Rmm CARTRIDGE SPECIFICATIONS

Cartridge Weight	278 gr. (18 gm.)
Cartridge Length	5.7 in. (145mm)
Case Length	39mm
Muzzle Velocity (in Air)	820 fps (250 mps)
Bullet Weight	203 gr. (13.2 gm.)

5.45x18mm MPTs (7N7) CARTRIDGE

The history of the MPTs cartridge is somewhat enigmatic. According to one Russian reference, the cartridge was individually developed by a woman, Antonina Deniskaya of TsNIITochmash. In another reference, credit is given to one Aleksandr Bochin. In yet another Russian reference, the designer of the cartridge is stated to be A.D. Denisova, another female designer. The Russians themselves seem to be somewhat ambivalent about the original designer of the cartridge. Regardless, the MPTs cartridge was originally developed for the PSM pistol (see Chapter 2), but has since been adopted in other pistols and has enjoyed some success as a self-defense load in Russia because the diminutive 5.45mm bullet has a relatively high sectional density, resulting in the ability to penetrate up to 45 layers of Kevlar, plus another 10 to 15mm of pine!

The MPTs' penetration capability came as a complete surprise to Western intelligence organizations when the PSM pistol was first test-fired in the late 1980s and early 1990s, but the terminal ballistics of the bullet could have been predicted had the bullet been analyzed using available computer models. With today's computer models, there should be no excuse for anyone's not having a fair idea of the ballistic performance of virtually any small-arms cartridge. The Russians emphasize the fact that the energy of the MPTs bullet is up to 1.5 times that of the .25 ACP (6.35mm Browning).

The MPTs cartridge was derived from the 9x18mm and is essentially a "necked-down" version of that cartridge. The statistics of the MPTs are not indicative of its overall terminal ballistics, despite the fact that the latter could have been derived from them. The latter is a function of the high sectional density of the 5.45mm bullet, which was obviously designed to achieve high levels of penetration. The bullet not only has a sharp ogive, it has a core that is divided into two portions. The front portion, which occupies approximately half the bullet, is steel; the rear is lead. There is a small air gap at the bullet tip forward of the steel penetrator. The cartridge remains popular in Russia for police use, and we can probably expect to see more small pistols chambered for it.

5.45x18mm MPTs (7N7) CARTRIDGE SPECIFICATIONS

Cartridge Weight	73.6 gr. (4.8 gm.)
Cartridge Length	.98 in. (25mm)
Case Length	18mm
Muzzle Velocity	1,033 fps (315 mps)
Bullet Weight	41.4 gr. (12.7 gm.)
Muzzle Energy	98 ft. lbs. (134J)

5.45x39mm CARTRIDGE

The history of the development of the 5.45x39mm was discussed in Chapter 1 under the section on the AN-94 assault rifle. The cartridge has been in Soviet and Russian service for more than 20 years and will serve into the foreseeable future; it has been adopted for the new Russian service rifle. However, as mentioned in Chapter 1, rifles have been developed in this caliber that use suppressors, necessitating subsonic ammunition for optimal effectiveness. As previously discussed, when a cartridge's velocity is reduced, it is necessary to increase the weight of the bullet to maintain terminal ballistics. That has been the case with the 7U1 cartridge developed for use in the AKS-74U. Bullet weight of the 7U1 has been increased to 78 grains (5.15 grams) fired from the AKS-74U at a nominal velocity of 994 fps (303 mps).

The second relatively new 5.45x39mm cartridge is the Model 7N10 high-penetration round. This cartridge is designed for use in standard AK-74- and AK-100-series rifles and fires a 55.5-grain (3.6-gram) bullet at a nominal velocity of 2,952 fps (900 mps).

Specifications of both these cartridges are shown along with those of the standard 5.45x39mm Model 7N6 round for comparison.

5.45x39mm CARTRIDGE SPECIFICATIONS

	Model 7N6	**Model 7N10**	**Model 7U1**
Cartridge Weight	161 gr. (10.5 gm.)	166 gr. (10.8 gm.)	188 gr. (12.25 gm.)
Cartridge Length	2.24 in. (57mm)	2.24 in. (57mm)	2.24 in. (57mm)
Case Length	39mm	39mm	39mm
Muzzle Velocity	2,952 fps (900 mps)	2,952 fps (900 mps)	994 fps (303 mps)
Bullet Weight	52.4 gr. (3.4 gm.)	55.5 gr. (3.6 gm.)	80 gr. (5.2 gm.)
Muzzle Energy	1,012 ft. lbs. (746J)	1,072 ft. lbs.	175 ft. lbs. (129J)

5.66x39mm MPS UNDERWATER RIFLE CARTRIDGE

The MPS underwater cartridge was developed in conjunction with the SPS pistol cartridge discussed above, and the principles behind its functioning underwater are identical to those of the pistol cartridge. The differences are in scale; the rifle cartridge is slightly larger in diameter and overall length. The case is rimless in order to function in the fully automatic APS underwater assault rifle. Given the larger case volume, the velocity of the 120mm-long MPS bullet is appreciably greater than that of the SPS pistol bullet, with concomitantly greater range both underwater and in open air. Effective range underwater is a maximum of 30 meters at 5 meters depth and 100 meters in open air.

5.66x39mm MPS SPECIFICATIONS

Cartridge Weight	398 gr. (26 gm.)
Cartridge Length	5.9 in. (150mm)
Case Length	39mm
Muzzle Velocity	1,197 fps (365 mps)
Bullet Weight	App. 284 gr. (18.5 gm.)
Muzzle Energy	788 ft. lbs. (581J)

6x49mm and 6x53mm CARTRIDGES

The 6x49mm cartridge was probably developed by TsNIITochmash for use in the 6mm "Unified" assault rifle and machine gun discussed above. There have been several proposals over the past 50 years asserting that a 6mm caliber would be the ideal military cartridge, striking a balance between the 7.62mm and 5.45/5.56mm, and although weapons in 6mm caliber have been experimented with in many nations, as of this writing the 6mm caliber has yet to find a home in any major military service.

There are a number of reasons for this, not the least of which is logistics. Most nations use basic infantry weapons in either NATO or Russian specification, which means that their weapons will be chambered for one of the standard calibers used by these countries. Adding a new cartridge in times of reduced military budgets would simply be too costly for the limited improvement that would be derived from changing over to the 6mm caliber.

This is probably one of the reasons that neither of these cartridges was ever adopted by Russia. Russia already has five standard military cartridges, and adding yet another would not only complicate ammunition logistics, but would add major production costs. That is probably why the 6mm weapons proposed by TsNIITochmash never stood a chance in competition against weapons chambered in standard Russian calibers.

Other than basic dimensions and a few references in periodicals, there have been no data whatsoever made public by the Russians regarding these experimental cartridges, which are included herein only for the sake of completeness.

7.62x35mm, 7.62x42mm, and 7.62x62.8mm SILENT CARTRIDGES

These three cartridges are considered together rather than separately because they are very closely related in concept, function, and the weapons that use them. In fact, two of them probably succeeded each other in Soviet and Russian military service.

The first cartridge is the 7.62x35mm cartridge, possibly designated the SP-2. This cartridge was used in the MSP pistol discussed earlier. The 7.62x35mm is virtually indistinguishable externally from the standard 7.62x39mm M1943 round. As can be seen in drawings, however, it is obvious that the internal workings of this cartridge are unusual indeed. There is a small propelling charge that drives two internal pistons forward to expel what is, to all appearances, a standard M1943 bullet out of the cartridge case. The first piston obturates against the cartridge case's internal shoulder while the smaller second one continues forward to expel the bullet and then obturates against the internal shoulder of the first piston, thereby retaining all the propelling gases and the associated sound.

There were several disadvantages to this system: its complexity, the fact that the protruding piston(s) complicated extraction and ejection, and that the cartridge case walls appear in photographs of sectioned cartridges to be no thicker than standard M1943 rounds. It is suspected that because of the thin walls, these cartridges experienced failures to retain the pistons, resulting in release of the propellant gases and the sound of the pistol firing. This suspicion is at least partially confirmed by the 7.62x62.8mm SP-3 cartridge fired from the S4M pistol, discussed in Chapter 2.

If the case walls of the first Russian silent cartridge were possibly too thin, those of the SP-3 that followed it went to the opposite extreme. The walls of this cartridge were greatly thickened and the design simplified somewhat in comparison with the earlier cartridge, but not in all ways. For example, the SP-3 cartridge has its own built-in firing pin; the hammer of the S4M pistol strikes the cartridge firing pin, which in turn strikes the primer, igniting the powder that sends the 122-grain bullet down the barrel of the S4M. The firing mechanism of the SP-3 cartridge is actually screwed into the base of the cartridge, an overly complex and expensive process. The bullet fired from the SP-3 is again a standard M1943 bullet.

Although I have never had the opportunity to check, it would be logical to assume that the rifling of both the MSP and S4M pistols is identical to that of an AK-47 or AKM rifle to further confuse matters when someone is assassinated by a person using either pistol. The scenario runs something like this: The "mark" is shot in the head on a crowded street using either an MSP or S4M. There is no discernible noise. Nobody hears any unusual sound, much less a shot. Yet when the autopsy is performed, there is an AK-47/AKM bullet! Where did it come from? How could someone have been killed on a crowded street in broad daylight with an AK and not a sound be heard? The reader can imagine the possibilities of using either of these pistols for assassinations, which is exactly what they were designed for. It is suspected that the S4M pistol and its SP3 cartridge are still in use today. They were used by the KGB on assassination missions in Central America during the 1980s. In fact, more than one was brought out of Central America, and at least one S4M is in the firearms "reference library" of a U.S. government agency.

The 7.62x42mm SP-4 cartridge is used in the PSS silent semiautomatic pistol. It is really of a more sophisticated design, but the PSS is not truly an assassination pistol in the sense that the S4M and MSP are. The PSS is obviously intended for special-operations use, where silence and quick follow-up shots are necessary, and the SP-4 cartridge design reflects that need. We know that it is not an assassination cartridge because the bullet it fires is absolutely unique—there is nothing like it in the world. The SP-4 bullet is nothing more than a mild-steel cylinder with a copper rotating band at the front to engage the rifling of the pistol. Moreover, the spent SP-4 casings are forcibly ejected from the pistol when fired. One does not leave spent cartridge casings lying about on assassination missions. Both of these characteristics indicate that the SP-4 is a special-operations cartridge for taking out sentries and for other missions where silence is desired for any of a number of operational reasons.

As mentioned, the SP-4 bullet is no more than a simple steel cylinder with a copper rotating band. The internal piston obturates against the relatively thick case wall, retaining the propellant gases while the bullet is launched down the barrel. The piston has a small nib in the center to help keep the bullet centered as it enters into the forcing cone of the barrel. Once it obturates against the internal shoulder of the cartridge case, the piston is nearly flush with the cartridge mouth, but does not protrude. This feature enables the cartridge to be used in a semiautomatic pistol rather than a manually operated one.

Needless to say, little has been made public regarding the 7.62x35mm and 7.62x62.8mm assassination cartridges, and they are little known in the West outside of intelligence communities, although some information has been published on them in the European firearms press. The pistols that fire these cartridges have never been test-fired in the West, nor have the Russians released any ballistic data on them. For that reason we must confine our specifications on these two cartridges to dimensional data. We do, however, have ballistic data on the SP-4 cartridge because the PSS pistol has been widely advertised for sale in the West. This is, of course, another suggestion that the S4M and MSP were intended for more clandestine purposes than the PSS.

7.62x35mm, 7.62x62.8mm, and 7.62x42mm SPECIFICATIONS

	7.62x35mm	**7.62x62.8mm**	**7.62x42mm**
Cartridge Length	52mm	App. 77mm	41.9mm
Case Length	35mm	62.8mm	41.5mm
Bullet Weight	122 gr. (8 gm.)	122 gr. (8 gm.)	143 gr. (9.9 gm.)

7.62x39mm M1943 CARTRIDGE

As I have mentioned, the Russians introduced a new variation of the venerable M1943 cartridge in the early 1990s. This new cartridge, designated the 57-N-231, retains the standard 122-grain (7.9-gram) weight of earlier versions of the M1943. The new bullet, however, has a steel core and, as mentioned, would be considered an armor-piercing round in any Western military force. The Russian military, however, categorizes the round as ball ammunition, and it is believed that the entire Russian ammunition industry has produced no other ball ammunition than this since 1991. According to Russian data, the 57-N-231 will penetrate a 10mm (13/32-inch) steel plate at a 100-meter range.

The Russians have also introduced a new subsonic cartridge, the 57-N-231U, for use in suppressed weapons, probably the OTs-14 and suppressed versions of the A-91, A-91M, AK-103, and AK-104 rifles and carbines. Like other subsonic rounds, the 57-N-231U has a relatively heavy bullet and reduced velocity in contrast to the standard round.

It is clear that the Russians have not abandoned the proven M1943 7.62x39mm cartridge, despite the introduction and standardization of newer cartridges. It seems, however, that M1943 ammunition will be for limited and specialized use. As we have seen, several new weapons have been developed in 7.62x39mm, but they are all special-purpose firearms, so domestic use of the cartridge will be limited. There is, however, a wide and probably steady demand for ammunition in this caliber, given the millions of AK-47 and AKM rifles in existence worldwide.

7.62x39mm AP AND SUBSONIC SPECIFICATIONS

	57-N-231 Armor Piercing	**57-N-231U Subsonic**
Cartridge Length	2.2 in. (56mm)	2.2 in. (56mm)
Case Length	39mm	39mm
Bullet Weight	122 gr. (7.9 gm.)	193 gr. (12.5 gm.)
Case Material	Steel	Steel
Bullet Material	Gilding Metal Jacket; Steel Core	Gilding Metal Jacket; Lead Core
Velocity from AKM	2,815 fps (715 mps)	968–1,017 fps (295–310 mps)

7.62x54Rmm CARTRIDGE

The Russian 7.62x54Rmm cartridge is the oldest standard military cartridge in the world, having been in continuous service since 1891 and showing no signs of retirement. The Russians usually refer to the cartridge as the 7.62x53mm, but as the case length is nominally 53.6mm and is rimmed, its usual Western designation is 7.62x54Rmm, which is used here. The Russian arms industry has introduced improved versions of this venerable cartridge, along with improved versions in other calibers. There are two M1891 cartridges worthy of note. The first is the high-penetration 7N13, and the other is the 7N14 sniper cartridge. According to Russian ammunition literature, the high-penetration cartridge is primarily intended for use in machine guns and the sniper cartridge in, naturally, SVD rifles. The 7N13 cartridge is unequivocally an armor-piercing round; the Russian specification tables for it state that it will penetrate a 10mm (13/32-inch) steel plate at a range of 400 meters. Penetration of soft body armor with this round is possible out to a range of 800 meters, although no specific type of body armor is stated. The 7N14 cartridge, on the other hand, is manufactured to a higher standard of precision for optimal accuracy from the SVD rifle and its variations, although this round may also be used in any weapon chambered in 7.62x54Rmm caliber. The bullet weight of the 7N14 sniper round is only slightly heavier than standard ball and the armor-piercing round, so rifling twist rate is not an issue with the 7N14 sniper cartridge.

7.62x54Rmm ARMOR-PIERCING AND SNIPER SPECIFICATIONS

	7N13 **Armor Piercing**	**7N14** **Sniper**
Cartridge Length	3.03 in. (77mm)	3.03 in. (77mm)
Case Length	53.6mm	53.6mm
Bullet Weight	145 gr. (9.41 gm.)	151.2 gr. (9.8 gm.)
Case Material	Steel	Gilding Metal
Bullet Material	Gilding Metal Jacket; Steel Core	Gilding Metal Jacket; Lead Core
Velocity from SVD	2,723 fps (820 mps)	2,723 fps (830 mps)

9x18mm MAKAROV, PMM, and SP-8

The original 9x18mm cartridge was developed as a replacement for the 7.62x25mm Russian cartridge. It was designed by Boris Semin, and the original bullet had a lead core in a steel-jacketed bullet, plated to reduce friction. Cases were brass. The latest conventional 9x18mm rounds have a gilding metal jacket with a mushroom-shaped steel core surrounded by lead. Recent cases are of lacquered steel.

The 9x18mm is the most powerful cartridge fired from a blowback-operated pistol, but it is decidedly marginal for military use, being far outclassed in terms of terminal ballistics by the 9x19mm (9mm Lüger or Parabellum) cartridge. Although the Russians never admitted to this fact, they tacitly acknowledged it when they introduced the Model 57-N-181M (PMM) cartridge, whose ballistics are squarely within the ballistic bounds of 9x19mm.

The new cartridge is, however, too powerful to be fired from a standard PM pistol. As previously mentioned, the Izmash engineers cleverly solved the problem of integrating the new cartridge into the Makarov pistol by putting spiral grooves into the chamber, which makes the pistol operate by delayed blowback, rather than straight blowback. The reason for this is that the cartridge case is held in place in the chamber by the spiral grooves by pressure until the bullet has left the muzzle, thereby allowing internal pressures and recoil forces to degrade to the point where they are manageable. Once the pressure drops, the case can be extracted and ejected.

The dimensions of the new cartridge and the older one are identical, so the PMM pistol can chamber and fire both types of ammunition, but the disadvantage of this is that so can older PM pistols! Chambering one of the new high-velocity cartridges in an older, straight-blowback-operated PM would likely be disastrous, so one can only wonder how the Russians manage to prevent this from happening. The 57-N-181M bullet can easily be identified by its conical ogive, but soldiers of all ranks cannot be counted on to follow instructions or ensure that they have the proper ammunition for their weapon. There are a number of potential solutions to this problem, but the Russians have made no mention of problems associated with new ammunition being fired in old pistols, so perhaps they have already resolved the issue.

A second, lesser known 9x18mm cartridge is the SP-8, designed for use in pressurized aircraft, indicating that the Russians may have a problem with aircraft hijacking. According to scanty Russian literature on this cartridge, it is intended for use at ranges of less than 25 meters and has the capability of "killing unarmored live targets," while at the same time not penetrating walls and aircraft fuselages.

Although the Russians have made very little data available regarding this specialized cartridge, it probably fires a frangible bullet that disintegrates upon impact with a hard surface. This is why it is ineffective against armored targets, even soft body armor. The SP-8 is a very specialized cartridge for use by special operations teams in antiterrorist operations on board aircraft and in other places where bullet penetration of walls and pipes might be a problem, such as nuclear facilities and densely populated areas subjected to terrorist attack.

9x18mm AMMUNITION SPECIFICATIONS

	57-N-181S Standard	57-N-181M High Penetration	SP-8 Frangible
Cartridge Length	.98 in. (25mm)	.98 in. (25mm)	.98 in. (25mm)
Case Length	18mm	18mm	18mm
Bullet Weight	91 gr. (5.9 gm.)	86.5 gr. (5.6 gm.)	77 gr. (5 gm.)*
Case Material	Gilding Metal	Gilding Metal	Gilding Metal
Bullet Material	Gilding Metal Jacket; Steel Core	Gilding Metal Jacket; Steel Core	Copper or Tungsten
Velocity	1,033 fps (315 mps)	1,345 fps (410 mps)	820 fps (250 mps)

*Estimated

9x21mm RUSSIAN

The RG-052/RG-054 9x21mm cartridge was first observed in 1993 when the Russian firm of TsNIITochmash announced the Gurza pistol, which is chambered for it. Unlike the 9x39mm cartridge described below, the Gurza is so far the only known weapon chambered for this cartridge. The 9x21 Russian is dimensionally similar to the Western 9x21mm cartridge, but is quite different in execution. The cartridge is based on the 9x18mm Makarov, not the 9x19mm. Case head dimensions of 9x21mm cartridges examined by me are identical to those of 9x18mm cases, so it is apparent that the new cartridge was derived from the 9x18mm. The 9x21mm, however, is a far more powerful cartridge than the Western 9x21mm. Its ballistics are in the lower end of .357 Magnum class, and 9x21mm Russian should never be chambered in a Western 9x21mm pistol, despite the fact that it probably will chamber.

Data on the RG-052 and RG-054 rounds have not been published, but both rounds are listed as having a "hard alloy core" bullet, and the RG-054 cartridges examined by me all have a steel penetrator whose nose is exposed at the tip of the bullet. From available photographs, the only apparent difference is that the RG-052 has a gilding metal case, whereas that of the RG-054 is lacquered steel. All available published performance data for the two cartridges are the same.

The 9x21mm Russian will probably never be widely distributed outside Russia, however, because it was designed to be an armor-piercing round. The nose of the bullet is formed by a steel penetrator, and the Russian designers make much of the fact that the 9x21mm Russian bullet will penetrate 30 layers of Kevlar and two 1.4mm-thick titanium plates at a 50-meter range. The pistol is also claimed to penetrate a 4mm steel plate at a range of 50 meters. During tests of this cartridge and the *Gurza* pistol in the United States, it was fired at NIJ Level IIIA (.44 Magnum protection) soft body armor and cleanly penetrated it.

According to TsNIITochmash representatives, the *Gurza* pistol is in use by MVD troops. It was developed to meet the rising threat of Russian criminals who can be counted on to be wearing body armor and be in large, heavy European automobiles.

There are a number of unanswered questions regarding this cartridge, not the least of which is why it has not appeared in any other of the many new pistols developed by the Russian firearms industry. The question also arises as to why the armor-piercing round seems to be the only one available. Having only one type of ammunition limits the usefulness and flexibility of the pistol and increases costs when expensive armor-piercing service ammunition must be used for training. Also, one cannot help but wonder why the Russians simply did not adopt the 9x19mm and develop a round similar to the Bofors HP 9x19mm armor-piercing cartridge, which achieves virtually the same terminal ballistics as the 9x21mm Russian.

9x21mm RUSSIAN RG-052/RG-054 SPECIFICATIONS

Cartridge Length	App. 1.38 in. (35mm) 5.9 in. (150mm)
Case Length	.82 in. (20.8mm)
Bullet Weight	103 gr. (6.7 gm.)
Case Material	Gilding Metal
Bullet Material	Gilding Metal Jacket; Steel Core
Velocity	1,410 fps (430 mps)

9x30mm *GROM* (THUNDER)

Little is known of the recently announced 9x30mm cartridge other than the fact that it is used in the new *Gepard* submachine gun (see Chapter 3). The *Grom*'s bullet leaves the muzzle of the weapon at appreciably higher velocity than its nearest Russian competitor, the 9x21mm RG-054. There appear to be two types of bullet for the *Grom*: ball and armor-piercing. The ball bullet appears identical to that of a 9x19mm, whereas the armor-piercing bullet seems to be the same as that fired from the RG-052/RG-054 9x21mm cartridges.

The cartridge itself appears to be yet another "stretched" 9x18mm, although this cannot be definitely ascertained without direct comparative measurement, which has not been accomplished as of the time this was written. Case material appears to be standard lacquered steel.

All in all, the 9x30mm cartridge appears to be a solution in search of a problem. Ballistically, it is only a marginal improvement over such existing cartridges as the 9x21mm Russian. Moreover, standard Russian intermediate caliber cartridges—such as the 5.45x45mm, 7.62x39mm, and 9x39mm fired from the AKS-74U, OTs-14, or A-91 compact assault rifles—achieve better ballistics than any pistol-class cartridge, whether the bullet from that cartridge is fired at 1,900 fps (600 mps), as is the 9x30mm, or 1,410 fps (430 mps), as is the 9x21mm Russian.

Unlike in Western countries, where ammunition developments usually are in response to perceived market demands, there is no similar firearms market per se in Russia. Thus, unless there is a military or police requirement, cartridges such as the 9x30mm are interesting technical exercises but probably will go the way of the 6mm cartridges discussed above: the Russian military and police already have a panoply of weapons that accomplish almost exactly the same purpose as the 9x30mm *Gepard* without having to worry about a new cartridge in the logistics inventory.

9x30mm *GROM* SPECIFICATIONS

Cartridge Length	App. 1.65 in. (42mm)
Case Length	30mm
Bullet Weight	103 gr. (6.7 gm.) (AP)*; 115 gr. (7.4 gm.) (Ball)*
Case Material	Lacquered Steel
Bullet Material	Gilding Metal; Lead Core (Ball) Gilding Metal; Steel Core (AP)
Velocity	1,968 fps (600 mps)

*Estimated

9x39mm SP-5 and SP-6

The 9x39mm cartridge was originally developed by TsNIITochmash for use in its AS and VSS suppressed rifles (see Chapter 1). The cartridge is now used in a number of special-purpose Russian weapons, which have been described above, and is even being considered by at least one U.S. manufacturer as a result of its overall effectiveness. There are two versions of the 9x39mm currently in production. The SP-5 is a ball round intended primarily for use in the VSS sniper rifle; the SP-6 armor-piercing round is intended primarily for use in the AS assault rifle. Both use the same principles to achieve terminal effects.

When a bullet is fired from a firearm, there are normally two elements to the sound that the gun makes: the blast associated with the expanding gases leaving the muzzle and a "crack" as the bullet leaves the muzzle and breaks the sound barrier, all of which are associated with cartridges that fire bullets at supersonic velocities. A suppressor (erroneously sometimes called a silencer) attenuates only the blast element of the noise. If the bullet leaves the barrel of the weapon at a supersonic velocity, there will still be a miniature "sonic boom" as it breaks the sound barrier leaving the muzzle. For this reason, most suppressed weapons use subsonic ammunition. The nominal velocity of sound at sea level is approximately 1,100 fps, depending on temperature, barometric pressure, and other variables, so subsonic cartridges are usually loaded to give velocities appreciably below this to accommodate for the variables, usually below 1,000 fps. However, when velocity is lost, so are range and energy. In a military cartridge, terminal effects are one of the primary performance criteria, so such a loss is undesirable. Partial accommodation can be realized by loading the cartridge with a heavier bullet, but if range and terminal effects are to be optimized, not only a heavier but also a larger diameter bullet is necessary. This usually raises both ballistic coefficient (which is a measure of the ability of a bullet to overcome air resistance and maintain velocity) and sectional density (which at its most simplistic is a measure of the ability of a bullet to maintain its velocity). Sectional density is a ratio of the bullet's diameter as compared with its weight. It is used to mathematically determine ballistic coefficient. It is not the purpose of this book to enter into an essay on ballistics, so I will leave the subject by noting that the SP-5 and SP-6 bullets have very high sectional densities and ballistic coefficients.

Because they have the characteristics mentioned above, these rounds are ideal for use not only in suppressed weapons but in compact assault rifles as well. These weapons are not intended to be used at ranges beyond 400 to 500 meters, and at these distances the SP-5 and SP-6 bullets have very high effectiveness. In fact, I witnessed a SP-6 bullet fired at a standard U.S. Kevlar helmet at 200 meters, and the SP-6 cleanly penetrated both sides. The Russian representatives claim that the bullet will penetrate both sides of such a helmet at 400 meters, and there is little reason to doubt their claim, given the ballistics and makeup of the rounds.

The 250-grain SP-5 ball round is of standard boat-tail configuration with a hollow base and steel-lead core. The rear half of the core is lead, and the forward half is steel with a small air space forward of the steel penetrator. The jacket is gilding metal with a lacquer seal at the case mouth and primer annulus. The case is lacquered steel for both rounds. The case itself appears to be no more than a 7.62x39mm case "necked-up" to accept the relatively large 9mm bullet. 9x39mm cases that the author has informally measured have essentially the same head diameter and case length as their smaller caliber cousins.

The 246-grain SP-6 is quite different from the ball round. The jacket is thin, and the core is entirely taken up by the mild-steel penetrator, except for a lead band surrounding the middle of the penetrator. It has been suggested to me that the band is present to allow for "crush" when the bullet is fired and engages the rifling of the rifle, because bore erosion with early armor-piercing ammunition of similar design was problematic. Whether this is the case remains to be seen, until such time as discussions can be held with TsNIITochmash regarding their bullet designs.

9x39mm SP-5 & SP-6 SPECIFICATIONS

	SP-5 Ball	**SP-6 AP**
Cartridge Length	2.2 in. (56mm)	2.2 in. (56mm)
Case Length	39mm	39mm
Bullet Weight	250 gr. (16.2 gm.)	246.2 gr. (15.9 gm.)
Case Material	Lacquered Steel	Lacquered Steel
Bullet Material	Gilding Metal Jacket; Steel-Lead Core	Gilding Metal Jacket; Mild-Steel Core w/Lead Band
Velocity	950 fps (290 mps)	950 fps (290 mps)

12.3x22Rmm, 12.3x40Rmm, 12.3x50Rmm, and 12.3x46Rmm CARTRIDGES

These cartridges are considered together because they were all, according to Russian literature, developed from 32-gauge shotgun shells. All were apparently developed independently with little consideration given to interchangeability, although this is difficult to imagine. As we have already considered, it is possible that the 12.3 cartridges were all developed together and were intended for use in both versions of the Udar revolver, with the Tula versions being able to chamber and fire only the shorter cartridges and the TsNIITochmash version having the ability to chamber and fire them all, which seems to be the case. If this was indeed the intent, one has to wonder why. Perhaps the Tula Udar will not take the pressures of the 12.3x50Rmm cartridge, or perhaps the cylinder will accept only a cartridge of 40mm maximum length. There are a number of different types of 12.3x40Rmm and 12.3x50Rmm cartridges, as can be seen from the accompanying tables. In addition to those listed in the table, there are tear gas and "flash-bang" 12.3x40Rmm cartridges. The 12.3x46Rmm is a shortened version of the 12.3x50Rmm for use with short-range plastic bullets.

The 12.5x50Rmm cartridge, so far as can be ascertained as of early 1998, was developed only for use in the OTs-20 Gnom smoothbore revolver. As with the 12.3mm cartridges, there are a number of different types of 12.5mm cartridges, including steel bullet, lead bullet, and shot. As of early 1998, there are no data available on the 12.5x50Rmm cartridges; until such time as data are forthcoming, information for 12.3mm cartridges are probably very close to 12.5mm performance levels.

At this point, rather than add any further to this already confusing situation, I will attempt to clarify matters by laying out each of the three calibers tabularly in hopes that the graphic presentation of these three cartridges will clarify the situation regarding them.

12.3x22Rmm SPECIFICATIONS

	Service Cartridge	Paint Bullet	Rubber Bullet
Cartridge Length	1.25 in. (32mm)*	1.25 in. (32mm)*	1.25 in. (32mm)*
Case Length	22mm	22mm	22mm
Bullet Weight	92 gr. 6 gm.)	47 gr. (3.1 gm.)	46.2 gr. (3 gm.)*
Case Material	Gilding Metal	Gilding Metal	Gilding Metal
Bullet Material	Steel	Plastic Paint	Rubber
Velocity	918 fps (280 mps)	512 fps (165 mps)	541 fps (165 mps)

*Estimated

12.3x40Rmm SPECIFICATIONS

	Steel Bullet	Shotshell	AP Bullet
Cartridge Length	2 in. (51mm)*	2 in. (51mm)*	2 in. (51mm)*
Case Length	40mm	40mm	40mm
Bullet Weight	139 gr. (9 gm.)	154 gr. (10 gm.)	108 gr. (7 gm.)
Case Material	Gilding Metal	Gilding Metal	Gilding Metal
Bullet Material	Steel	Lead	Steel*
Velocity	934 fps (285 mps)	984 fps (300 mps)	1,148 fps (350 mps)

*Estimated

12.3x50Rmm & 12.3x46Rmm SPECIFICATIONS

	"Live" Cartridge	Plastic Bullet
Cartridge Length	1.96 in. (50mm)	1.92 in. (49mm)
Case Length	50mm	46mm
Bullet Weight	154 gr. (10 gm.)*	47 gr. (3.1 gm.)*
Case Material	Gilding Metal	Gilding Metal
Bullet Material	Lead	Plastic
Velocity	984 fps (300 mps)*	541 fps (165 mps)*

*Estimated